Neural Network Analysis, Architectures and Applications

Related titles published by
Institute of Physics Publishing

Neural Computing: An Introduction
R Beale and T Jackson

Handbook of Neural Computation
Edited by E Fiesler and R Beale

*Neural Network Perspectives on Cognition
and Adaptive Robotics*
Edited by A Browne

Neural Network Analysis, Architectures and Applications

Edited by

Antony Browne

Nene College, UK

CRC Press
Taylor & Francis Group
Boca Raton London New York

CRC Press is an imprint of the
Taylor & Francis Group, an **informa** business

First published 1997 by IOP Publishing

Published 2019 by CRC Press
Taylor & Francis Group
6000 Broken Sound Parkway NW, Suite 300
Boca Raton, FL 33487-2742

ISBN 13: 978-0-7503-0499-3 (hbk)

Visit the Taylor & Francis Web site at
http://www.taylorandfrancis.com

and the CRC Press Web site at
http://www.crcpress.com

British Library Cataloguing-in-Publication Data

A catalogue record for this book is available from the British Library.

Library of Congress Cataloging-in-Publication Data are available

Typeset in TEX using the IOP Bookmaker Macros

To Jayne and Liam

Contents

Preface

In the past few years there has been an explosion of interest in neural networks. This interest is in both the theoretical implications of neural network systems and the commercial exploitation of neural computing solutions by business and industry. Many people are interested in neural networks from many different perspectives. Cognitive scientists use neural networks to build models of human cognition. Engineers use them to build practical systems to solve commercial and industrial problems. The number of people involved in the study or use of neural network systems grows daily.

In the past a neural network was often viewed as a 'black box' solution to a problem—it worked but there was little understanding of how it worked. Recently techniques have been developed that extract information from a trained network that represents what the network has actually learned. Part of this book describes methods of extracting information from trained networks and also discusses the important topic of network simplification.

Most people using neural computing techniques are familiar with simple feed-forward networks, modelled in software using the backpropagation algorithm. More recently, new hardware architectures and training algorithms have been developed to increase the speed of neural networks in both their training phase and final application. In part of this book hardware architectures are discussed, together with a description of fast training algorithms for feed-forward networks.

There is a huge number of commercial and industrial uses of neural networks. They are used in fields as diverse as medical diagnosis, direct marketing, financial forecasting and the control of industrial processes. It would be impossible to give a comprehensive overview of all neural networks applications in one book. However, the final part of this book contains three interesting and diverse applications.

This book is designed to give a perspective on some of the many areas of neural computing. It is not written as an introductory textbook; it is assumed that the reader has some previous knowledge of neural networks and an understanding of their basic mechanisms. The book is divided into several parts, each of which attempts to give a brief overview of an area of neural computing. These parts are:

- *Understanding and simplifying networks.*   Methods for extracting information about what a trained neural network has learned are outlined, together with a method for simplifying network architectures based on information theory.
- *Novel architectures and algorithms.* Two novel hardware implementations of neural networks are described, together with a discussion of fast training algorithms for feed-forward network architectures, and a discussion of the use of multiple feed-forward networks in parallel.
- *Applications.* Some applications of neural networks in the diverse fields of control (including neuro-fuzzy control), data compression and the use of recurrent networks for target identification are discussed.

Many chapters in the book are cross-referenced to a companion volume *Neural Network Perspectives on Cognition and Autonomous Robotics* (1997, Institute of Physics Publishing), which can be seen as complementary to this volume as it contains chapters on:

- *Representation.* This part delves into questions on the representational power of neural networks, and discusses the use of information theory in neural computing. These parts are:
- *Cognitive modelling.* The use of neural networks for cognitive modelling is discussed, including both the modelling of human reasoning and the implications of neurophysiological data.
- *Adaptive robotics.*   Robots which can adapt to their environment are described, together with a discussion on biologically realistic learning mechanisms.

Hopefully the following chapters will clarify some of the many fields in which neural computing is expanding. No textbook can ever hope to give a comprehensive review of the myriad directions in which current research is headed. However, this volume attempts to give a flavour of some of the most promising areas.

A Browne
July 1997

PART 1

UNDERSTANDING AND SIMPLIFYING NETWORKS

Gone are the days when a neural network was viewed as a 'black box' solution to a problem. More and more techniques are being developed to extract information from a trained network representing what that network has actually learnt. Such techniques are important, not just in understanding how the network is performing its task, but also when networks are used in safety-critical applications, where a guarantee of the networks' performance under all situations is essential. This part of the book describes methods of extracting information from a trained network, and also describes a method which uses information theory to simplify the architecture of networks. This architectural simplification also helps to make networks more understandable.

Chapter 1

Analyzing the Internal Representations of Trained Neural Networks

J A Bullinaria
Birkbeck College, UK
j.bullinaria@psyc.bbk.ac.uk

1.1 Introduction

Although it may sometimes be sufficient to view a trained neural network as a 'black box' that solves or performs a particular task, we would often like to have some idea of what exactly the network is doing. It is clearly something of a problem not to understand how a network operates when it has been designed as a model of some cognitive function. It is equally problematic to have artificial systems solving real world problems in ways that are not easily understandable by their human creators. Such problems range from matters of legality (e.g. in some countries it is illegal to refuse credit without giving a reason and 'because our neural network said so' is not generally considered to be a good enough reason) to matters of reliability (e.g. in safety critical applications, can we trust a system whose working we do not fully understand?).

In this chapter we shall be concerned with analyzing the internal representations that are learnt by simple feedforward networks and using the resultant information to understand how the networks are operating. This will also allow us to investigate how the networks are likely to respond to damage, which is important both from the point of view of neuropsychological modeling and for the more practical problem of estimating the robustness of implemented real world systems [164].

We shall begin by reviewing the traditional techniques of hierarchical cluster analysis, principal component analysis, multi-dimensional scaling and discriminant analysis and illustrate their use by investigating the internal representations learnt by a recent connectionist model of reading aloud. We shall see that there are limitations to all these approaches and then show how simple output weight projections may lead to a clearer picture of what is happening. In the reading model, the learning trajectories of these projections may help us understand reading development in children and the results of naming latency experiments in adults. We shall then discuss contribution analysis and its use in predicting the effect of various types of damage to the networks. In our reading model, studying the effects of network damage seems to provide insight into the mechanisms underlying acquired surface dyslexia.

Before embarking on our general analysis of internal network representations, we first describe the network model that we shall use to illustrate our analysis. The NETtalk [204] model of reading aloud (i.e., text to phoneme conversion) was one of the earliest applications made possible by the re-invention of training techniques suitable for multi-layer neural networks [80]. It has recently been shown [11] how this original NETtalk model can be modified to work without the need for pre-processing of the training data to align the letters and phonemes prior to training. This modified model not only has superior learning and generalization performance to earlier reading models trained on the same words [83], but also has the advantage that it does not require the use of complicated input and output representations. Consequently, it has become feasible to analyze the internal representations of this model with a view to better understanding how it operates under normal conditions and after damage.

There are numerous possible variations of the original NETtalk model [9]. We shall only be concerned here with a fairly standard version consisting of a fully connected simple feedforward network with sigmoidal activation functions and one hidden layer of 300 units, as shown in figure 1.1.

The input layer consists of a window of 13 sets of units, each set having one unit for each letter occurring in the training data (i.e., 26 for English). The output layer consists of two sets of units, each set having one unit for each phoneme occurring in the training data (i.e., 38 units). The network was trained using back-propagation on a standard set of 2998 monosyllabic words with the corresponding pronunciations.[1] The input words slide through the input window, starting with the first letter of the word at the central position of the window and ending with the final letter of the word at the central position, with each letter activating a single input unit. The output phonemes correspond to the letter in the centre of the input window in the context of the other letters in the input window. Usually the output consists of one phoneme and one phonemic null (e.g., 't' $\rightarrow$ /t_/ in 'hot'), occasionally it consists of two phonemes (e.g., 'x' $\rightarrow$ /ks/ in 'box') and for silent letters we get two phonemic nulls (e.g., 'e' $\rightarrow$ /_ _/ in

---

[1] We use the phonemic notation and conventions and words of Seidenberg and McClelland [83] throughout this chapter. Apart from the standard consonants, this has: D = 'th' in 'than', T = 'th' in 'thin', S = 'sh' in 'shot', C = 'ch' in 'chat', N = 'ng' in 'fang', Z = 'z' in 'azure', a = 'a' in 'hat', A = 'ai' in 'mail', e = 'e' in 'met', E = 'ee' in 'deed', i = 'i' in 'hit', I = 'i' in 'pint', o = 'o' in 'hot', O = 'oa' in 'goal', ^ = 'u' in 'hut', U = 'oo' in 'boot', u = 'oo' in 'book', * = 'aw' in 'saw', W = 'ow' in 'cow'.

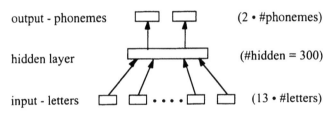

output - phonemes   (2 • #phonemes)

hidden layer   (#hidden = 300)

input - letters   (13 • #letters)

Figure 1.1. The NETtalk style model we will use to illustrate our discussion.

'cake'). The fact that we have three possibilities causes the so-called *alignment problem*, because it is not obvious from individual words in the training data how the letters and phonemes should line up. The advantage of this model over the original NETtalk is that, rather than doing the alignment by hand prior to training, a multi-target approach [9] allows the network to *learn* the appropriate alignments during the training process. Given a word such as 'huge' → /hyUdZ/, the network considers all possible output target alignments (e.g., /hy Ud _ _ Z_/) and trains only on the one that already gives the smallest total output activation error. Even if we start from random weights, given a sufficiently representative training set, the sensible regular alignments will tend to over-power the others, so the network eventually settles down to using only the optimal set of alignments (e.g., /hy U_ dZ _ _/). Once trained, this network achieves perfect performance on the training data (including many irregular words) and 98.8% on a standard set of 166 non-words used to test generalization. It also provides simulated reaction times that correlate well with various naming latency experiments and allows several possible accounts of developmental and acquired surface dyslexia. Most important for current purposes, however, is that we have a simple network successfully trained to perform a relatively complex mapping.

1.2 Analysis techniques

In the reading model, different regions of hidden unit activation space are selected by the output weights to activate different output phonemes. The learning process consists of judiciously choosing these regions and mapping from each central input letter to an appropriate region depending on the context information (i.e., surrounding letters). Since the consistent weight changes corresponding to regularities will tend to reinforce, whereas others will tend to cancel, the network tends to learn the most regular mapping possible and hence we also get good generalization performance. This manifests itself here, and in other multi-layer feedforward networks, in the formation of well structured internal representations, i.e., patterns of hidden unit activation. Each network input pattern maps onto a particular point in hidden unit activation space and by understanding how these points stand in relation to each other we can gain an understanding of how the network is operating. In this section we shall critically review several techniques that have previously been employed to study the internal representations learnt by connectionist systems and illustrate their use with our reading model. We shall also look at some less traditional techniques which appear to cast more light on what is happening.

1.2.1   Hierarchical cluster analysis

One way to map out what is going on in hidden layer activation space is to perform a hierarchical cluster analysis (HCA) of the points corresponding to each input pattern [30]. The basic idea of HCA is that we define some distance measure on the hidden layer activation space (such as simple Euclidean distance) and then construct a hierarchy of clusters of points based on that measure. If the network is operating efficiently, we can expect input–output patterns to be more closely clustered when they are more related. This approach was found to result in sensible clustering of the letter-to-phoneme correspondences in the original NETtalk model [204] and also of the lexical categories in a simple sentence prediction network [28]. We might therefore expect this approach to identify useful relations in other situations where they might not be so obvious.

Rather than attempting to look at all the 12 744 points representing the training data of our reading model (given by 2998 words with an average of 4.25 letters per word), we begin by looking at the mean activations for each of the main 65 letter-to-phoneme mappings. A simple Euclidean clustering results in figure 1.2 and we obtain a similar picture using an L1 norm.

The overall pattern is largely as one might expect: vowels together, silent letters together, consonants together and so on down to the likes of /dZ/ sounds together; though there are a few anomalies (such as 'k' → /k/ being grouped with the silent letters) that we shall discuss later. It is also worth noting that the words are sometimes initially clustered according to their input letters (e.g., the 'a' → /o/ instances are clustered with the other 'a' instances rather than the 'o' → /o/ instances) and sometimes according to their output phonemes (e.g., the 'i' and 'y' words are clustered according to the output /i/ and /I/ sounds).

Such large-scale clustering is interesting, but we also need to check that the good clustering persists right down to the level of individual words. To illustrate this, figure 1.3 shows the clustering of a representative set of 72 instances of the single vowel 'i' (and the words containing them). We see that there appears to be a clear distinction between the long /I/ sound and the short /i/ sound. However, a closer inspection shows that the irregular words (such as 'give' → /giv/ and 'pint' → /pInt/) are clustered with their regular counterparts ('gibe' → /dZIb/ and 'tint' → /tint/) rather than with the other words pronounced in the same way. Also, we find whole sub-rules (e.g., '–ind' → /–Ind/) apparently in the wrong high-level cluster. It seems that the HCA is representing the well known linguistic rule that a final 'e' lengthens the preceding vowel, but it is not picking up the fact that the network has also managed to learn the exceptions to that rule. In many situations the whole object is for the network to identify

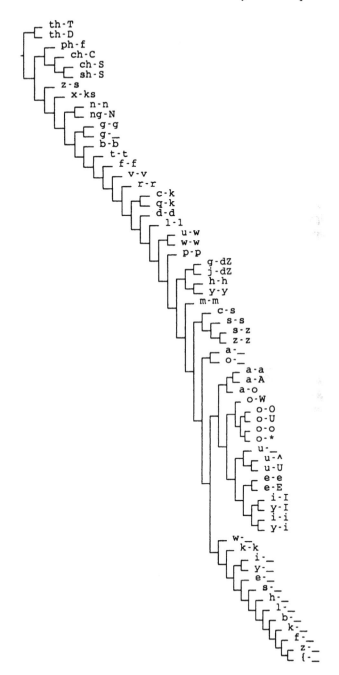

Figure 1.2. HCA of the mean letter–phoneme points in hidden unit space.

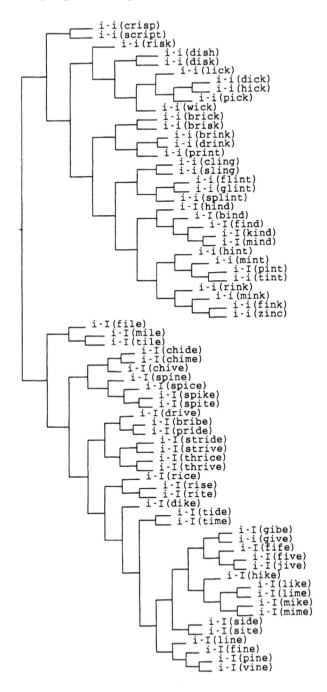

Figure 1.3. HCA of representative 'i' words in hidden unit space.

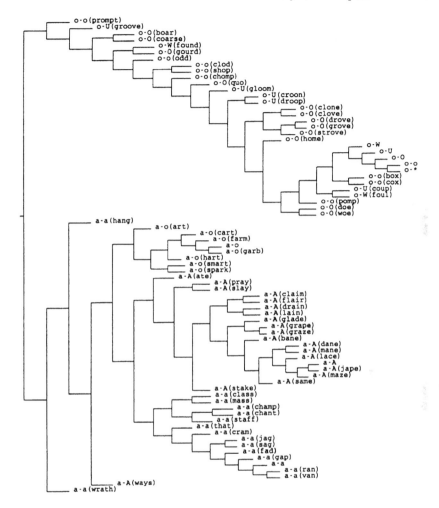

Figure 1.4. HCA of representative 'a' and 'o' words and their letter–phoneme means.

any regularities in the training data and to ignore the exceptions (that commonly constitute 'noise'), so HCA may well still be a useful tool in these circumstances, but it gives a misleading picture of the network's performance.

Another potential problem we face is that some parts of the training data may cluster better than others. Figure 1.4 shows the cluster plot for 68 instances of the letters 'o' or 'a' in our reading model plus the relevant eight mean mapping points from figure 1.2. We have a clear distinction between the 'o' and 'a' words, and the 'a' words split reasonably well into the /o/, /A/ and /a/ clusters with the mean values fairly central to each cluster. The 'o' words, however, show poor clustering with the means apparently closer to each other than they are to the words they represent.

The above examples illustrate what the HCA can miss and consequently, even when the clustering appears to make sense, we must be careful about what conclusions we draw from such an analysis. Since the network itself does not make use of Euclidean (or other) distance measures on the hidden unit activation space, it is not surprising that HCA can sometimes produce slightly misleading results. Indeed, all the inter-point distances are actually very similar in the examples shown above (mean 4.0, s.d. 0.6 for figure 1.2; mean 3.8, s.d. 0.8 for figure 1.3; mean 4.4, s.d. 0.9 for figure 1.4), so clustering does not make much sense anyway. Given a large enough dimensional space, the points will tend to spread themselves out as uniformly as they can. Reducing the number of hidden units closer to the minimum number required to learn the mapping may help slightly, but even if we train our reading network with only 30 hidden units we still find the exception words falling in the wrong clusters.

Since the network's output weights operate by projecting out particular subspaces of the hidden unit activation space, to get a better understanding of the internal representations we really need to see more directly how the words are positioned in the hidden unit activation space.

1.2.2   Principal component analysis

For the reading model we deliberately chose to use a large number of hidden units (i.e., 300), about ten times as many as actually needed to learn the training data. The reason for this was that we were particularly interested in modeling the effects of brain damage and acquired dyslexia. To do this realistically we needed a system that was fairly resilient and degraded gracefully when damaged. This required a highly distributed internal representation for which the removal of any single hidden unit or connection had very little effect on the network's performance.

We succeeded in this aim, but are now left facing the difficult problem of visualizing points in a 300-dimensional space. Even if we had used a more minimal network, with only around 30 hidden units, we would still have far too many dimensions to visualize easily. We clearly need to reduce the number of dimensions to something more manageable, i.e., two or three. One conventional way to do this is to use principal component analysis (PCA). If $\{P_{i\alpha} : i = 1, \ldots, d; \ \alpha = 1, \ldots, n\}$ are the vector components of a set of n points in our d-dimensional hidden unit activation space and $\langle P_i \rangle$ denotes the mean $P_{i\alpha}$ over all values of α, then equation (1.2) defines the standard covariance matrix S_{ij}. It follows that, since S_{ij} is symmetric, the matrix Λ_{ij} of its eigenvectors given by equation (1.1) is orthogonal.

$$\sum_k S_{jk}\Lambda_{ki} = \lambda_i \Lambda_{ji} \tag{1.1}$$

$$S_{ij} = \sum_{\alpha} (P_{i\alpha} - \langle P_i \rangle)(P_{j\alpha} - \langle P_j \rangle). \tag{1.2}$$

This means that Λ_{ij} can be used to perform a change of basis, i.e., an axis rotation, as given by

$$P^{\Lambda}_{i\alpha} = \sum_{j} \Lambda^{-1}_{ij} P_{j\alpha} \tag{1.3}$$

such that the covariance matrix is diagonalized as in

$$S^{\Lambda}_{il} = \sum_{j} \sum_{k} \Lambda^{-1}_{ij} S_{jk} \Lambda_{kl} = \lambda_i I_{il} \tag{1.4}$$

and the total variance is unchanged and given by

$$\text{var}(P) = \text{trace}(S) = \text{trace}(\Lambda S^{\Lambda} \Lambda^{-1}) = \text{trace}(S^{\Lambda}) = \sum_{i} \lambda_i. \tag{1.5}$$

The p new coordinates $\{P^{\Lambda}_{i\alpha} : i = 1, \ldots, p; \ \alpha = 1, \ldots, n\}$ corresponding to the p largest eigenvalues λ_i are called the first p principal components and provide the best possible account for the variance in p dimensions.

PCA thus provides a convenient procedure for dimensional reduction with the minimum loss of information. We simply project our points onto the p-dimensional sub-space spanned by the first p eigenvectors of the covariance matrix. This approach was used to good effect in [29] to analyze a sentence processing network. We see in figure 1.5 that the first two principal components alone are able to separate the vowels, consonants and silent letters in our reading model. However, at the level of individual words, we see in figure 1.6 that many exception words (e.g., 'give') and sub-rules (e.g., the '-ind' words) still find themselves clustered incorrectly.

The problem is that, in our network, the variance is distributed over too many components (the first three normalized eigenvalues for the full set of training data are 0.096, 0.078, 0.067). Moreover, the situation is only slightly improved if our network has only 30 hidden units (eigenvalues 0.174, 0.117, 0.086). Clearly, taking only the first two or three components on their own is bound to give a very poor representation of what is happening.

1.2.3 Multi-dimensional scaling

A useful non-metric approach to dimensional reduction is provided by multi-dimensional scaling (MDS). A gradient descent algorithm is used to adjust iteratively the positions of the points in a low-dimensional space until the rank order of the inter-point distances corresponds as closely as possible to those in the original space [116]. Since we can start this iterative procedure with the positions given by PCA, we are guaranteed to end up with at least as good a representation of the inter-point distances as provided by PCA on its own,

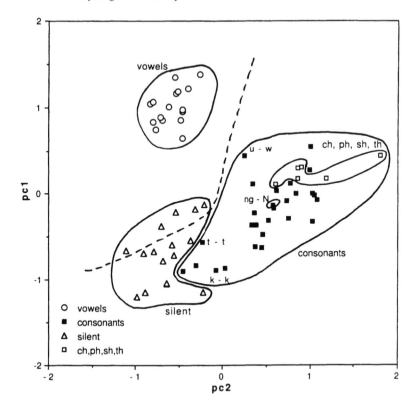

Figure 1.5. The first two principal components of the mean letter–phoneme positions.

though in practice, since the procedure can settle into local minima rather than the optimal configuration, it is usually sensible to start the procedure from a number of different configurations and select the best final configuration.

For small numbers of points, MDS works quite well. The average phoneme data of figures 1.2 and 1.5 result in the two-dimensional MDS plot shown in figure 1.7.

The rank correlation with distances in the original data is 0.82, compared with 0.50 for a 1D plot, 0.88 for a 3D plot and 0.56 for the first two principal components. In addition to the vowel, consonant and silent clusters evident in the PCA plot, a 'ch, ph, sh, th' cluster has separated itself as it did in the HCA. We can not only now see more clearly how the points are clustered but also better understand the anomalies in the HCA, e.g., why the 'k–k' point was grouped with the silent letters. Figure 1.8 shows that we can also get good plots for the individual words. We still have problems with the exception words and sub-rules, but there is a stronger tendency for them to appear at the edges of clusters as close as possible to the clusters of their regular counterparts. The

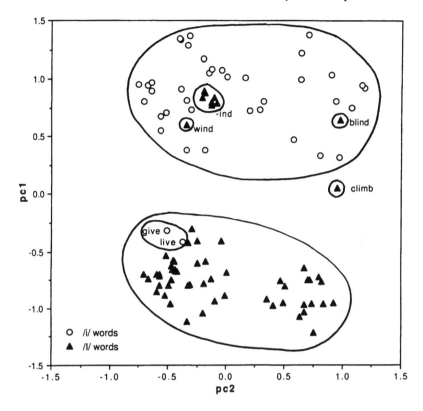

Figure 1.6. The first two principal components of our typical word set.

rank correlation with the original data here is now 0.90 compared with 0.83 for the PCA. However, for larger numbers of points the correlations become weaker and often words that we know from cluster analysis should be close together do not appear together on the MDS plots. In these cases it is clearly dangerous to make detailed predictions from MDS plots, since it is not clear which lost information is responsible for the breakdown in correlation.

1.2.4 Discriminant analysis

The problem with both PCA and MDS is that they fail to take into account the fact that some hidden units are more important than others, i.e., they do not take into account the network's output weights. It is also becoming clear that we cannot expect to represent reliably the whole of our network's internal representation in only two dimensions. What should be feasible and more useful, however, is to plot a series of small useful subsets of the full representation based on the knowledge that we already have about the network's operation. For example, we could attempt to elucidate the network's distinction between

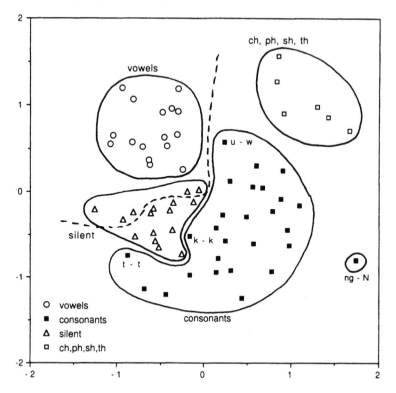

Figure 1.7. A two-dimensional MDS plot of the mean letter–phoneme positions.

the long /I/ and short /i/ sounds. We can do such a thing using discriminant analysis, which is a general procedure for projecting onto sub-spaces that optimize particular conditions [25]. This approach, in the form of canonical discriminant analysis (CDA), was successfully applied to study combinatorial structure in hidden unit space [106]. We shall attempt to use it to identify the hidden unit sub-spaces responsible for particular output patterns in our reading model.

If we know which of G groups each point in hidden unit space belongs to (e.g., which output phoneme it corresponds to), we can partition the total covariance matrix $S = W + B$ into the within groups covariance given by

$$W_{ij} = \sum_g \left(\sum_{\alpha \in g} (P_{i\alpha} - \langle P_i \rangle_g)(P_{j\alpha} - \langle P_j \rangle_g) \right) \qquad (1.6)$$

and the between groups covariance given by

$$B_{ij} = \sum_g n_g (\langle P_i \rangle_g - \langle P_i \rangle)(\langle P_j \rangle_g - \langle P_j \rangle) \qquad (1.7)$$

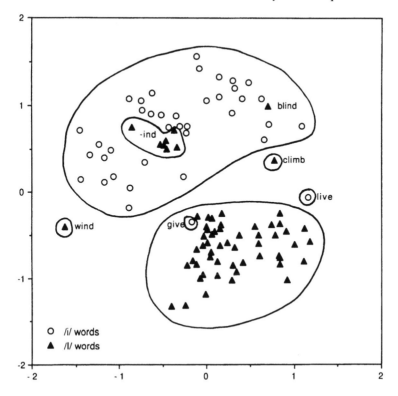

Figure 1.8. A two-dimensional MDS plot of our typical word set.

where the groups are labeled by g, contain n_g points and have mean components $\langle P_i \rangle_g$. The aim of discriminant analysis is to find a low-dimensional sub-space which best discriminates between the given groups. Since, roughly speaking, the determinant of a covariance matrix is a measure of the dispersion, a convenient fitness function to maximize is the ratio $|B|/|S|$ since it is well known [43] that this maximization can be achieved by solving the eigenvalue problem for $S^{-1}B$ to give a matrix M which projects our points onto a (rank(B) $\leq G - 1$)-dimensional sub-space. In this sub-space the points will be clustered into groups with the maximum between group separations and minimum within groups dispersion. Unlike with PCA, the projection directions are not necessarily orthogonal, but still their eigenvalues provide a useful measure of the importance of each direction.

Since we know that our network performs a similar clustering [33], it is tempting to assume that this procedure will give a good representation of what is happening in hidden unit space. Consider our long /I/ versus short /i/ case again. We can separate the words into two groups and use CDA to obtain a projection vector in hidden unit space that best discriminates between the two 'i' sounds. The quality of the discrimination will clearly depend on the number of data points we use. If we have many fewer points than the number of hidden units, then the discrimination is essentially perfect ($B/S = 1.0000$ for 160 points). In fact, we can get equally good discrimination even if we assign the points to groups at random ($B/S = 1.0000$). It is clear that this is not giving us a good picture of the true internal representation. If we use all the points in the training data (239 /I/s and 272 /i/s) we do better. We then obtain $B/S\hat{E} = 0.98$ for the true groups and $B/S = 0.61$ for random groups and, as we should expect, for random groups we fail to get good clusters at all and have many overlaps. However, if we test this procedure on words or non-words not in the training data, the projection vector fails to classify them properly even when the network itself does. To define the projection more accurately we clearly need many more data points, particularly for the borderline region between the two groups. To this end a set of 14 766 words and non-words of the form '$C_1 V C_2$' and '$C_1 V C_2 e$' was generated, where C_1 was one of a set of 58 initial consonant clusters, V was one of the set {i, ia, ie, y} and C_2 was one of a set of 58 final consonant clusters. Using these, the CDA then gave us a projection with $B/S = 0.83$, but there was now a large overlap between the two groups: $\max_i = -0.15, \min_i = -0.38, \max_I = -0.23, \min_I = -0.46$ with 2454 /I/ words greater than $\min_i$ and 3828 /i/ words less than $\max_I$. Figure 1.9 illustrates the problem with a more manageable intermediate set of 942 words and non-words. As usual, the exception words (e.g., 'give' and 'pint') are at the forefront of the problems.

It is clear that standard CDA does not necessarily give us a good representation of the true internal representation. The problem is that, when the network learns, it certainly maximizes the between group distances $\min_i - \max_I$, but it has no need to minimize the within group dispersions. This is simply because, for our non-linear networks, once the projections fall in the tails of the output sigmoids, very large differences can make relatively little difference to the networks' actual outputs [33]. One way we may attempt to get round the limitations of standard discriminant analysis is to start with CDA and then employ a simple iterative procedure to adjust the projection vectors so that all the points are correctly classified. There are many ways we can do this. Suppose, for example, $P_{i\alpha}$ is a point in hidden unit activation space and V_i is our projection vector, then the projection of each point is given by

$$D_\alpha = \sum_i V_i P_{i\alpha}. \tag{1.8}$$

This projection can be increased or decreased by a standard gradient descent adjustment of V_i given by

$$\Delta V_i = \pm\varepsilon \frac{\partial D_\alpha}{\partial V_i} = \pm\varepsilon P_{i\alpha} \tag{1.9}$$

where ε is a small constant. If we keep ε sufficiently small and sum the ΔV_i

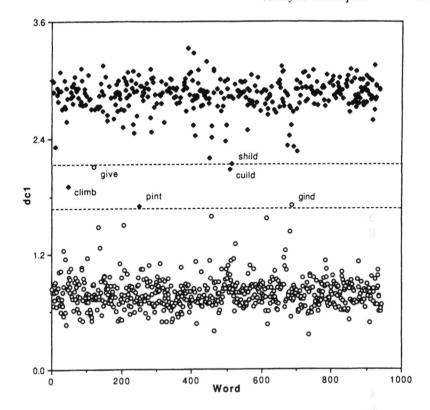

Figure 1.9. The canonical discriminant components for the /i/–/I/ distinction.

over an appropriate subset of points (e.g., all the misclassified points) we can iteratively decrease the overlap.

When this was done for our 14 766 data points in the reading model it resulted in the correct classifications with a reduced $B/S = 0.73$. Unfortunately this was still not good enough. For a good representation, we would expect the borderline cases in the projections to correspond to borderline output phonemes—in fact, the correlation was very poor. Moreover, the same iterative procedure even managed to find a projection vector that could classify the set of data points into randomly assigned groups ($B/S = 0.49$).

Projection vectors provided by discriminant analysis are clearly not very useful if they can be found for groupings that bear little or no relation to what the network has actually learnt to do. Such spurious projections are possible because the procedure can make use of any noise that results from having a large number of extra hidden units that are not strictly necessary for performing the mapping (about 270 in the case of our reading model). If our network has very nearly the minimal number of hidden units (i.e., 30), the CDA still gives a projection vector that has the two groups overlapping ($B/S = 0.75$) and it is still possible to adjust the vector to separate the groups ($B/S = 0.69$). However, it is no longer possible to find a projection vector that correctly classifies the random groups.

We can conclude, therefore, that in general standard CDA does not reliably inform us how a network is operating. Using gradient descent procedures to find projection vectors that separate the groups can also lead to misleading results. The problem is that we can very easily end up with projection vectors that have little to do with the actual network outputs. Networks that have many hidden units more than actually required for the task in question will be particularly susceptible to these problems. At least, by attempting to use the networks' hidden unit activations to classify the points into random groups, it is possible to get some estimate of the reliability of the projection vectors in particular situations.

1.2.5 Output weight projections

For the simple one-hidden-layer architecture and localist output representation of our reading model, it is actually very easy to find projection vectors that correlate with the outputs. We can simply use the projections the network itself has learnt—namely the output weights. If we project the hidden unit activations using the output weights W_{ij} and redefine the zero points using the output thresholds θ_j, our projections are then simply the network outputs before being passed through the sigmoid. We are guaranteed rank correlation. We thus have a suitable projection vector for each phoneme and these 38 vectors actually turn out to be nearly orthogonal (mean angle 84°, s.d. 5°). These projections can then be viewed in pairs to examine the relationships between the clusters, or any of the above techniques may be used to study interesting sub-spaces of this 38-dimensional space (e.g., the four-dimensional /i/, /I/, /e/, /E/ sub-space may be studied to investigate the various pronunciations of 'ie'). Figure 1.10 shows the resultant discrimination for our /I/ and /i/ phonemes for our 300-hidden-unit network.

Each point has a positive projection onto the line in hidden unit space corresponding to that phoneme and a negative projection onto the lines corresponding to all the other phonemes. Thus points corresponding to other phonemes would appear in the bottom left quadrant.

The projection vector that we were attempting to find by discriminant analysis now corresponds to the /i/–/I/ diagonal in figure 1.10. We can easily see the clear separation of the groups and the unrestricted within group dispersion that really corresponds to what the network has learnt ($B/S = 0.72$). We can also easily check the relation of these correct projection vectors to those provided by the CDA and gradient descent group separation. For the 300-hidden-unit network the angles between the vectors are 113° and 63° respectively. For the 30-hidden-unit case the corresponding angles are 24° and 3°, which is consistent

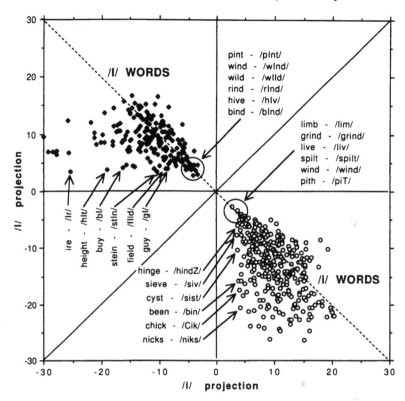

Figure 1.10. The 2D hidden unit sub-space corresponding to the /i/ and /I/ phonemes.

with our intuition that we get more reliable results for networks with fewer spare hidden units.

Plotting trajectories on graphs such as figure 1.10 can help us understand how the final pattern of projections arises, as well as possible causes of developmental problems (such as developmental dyslexia) and how the network responds to damage. Of course, the output weights and thresholds change during the learning or damage process at the same time as the hidden unit patterns, so whilst the following description is broadly correct, we should be careful not to take it too literally.

If we begin training with small random initial weights, all hidden unit activation points start near coordinates A_j given by

$$A_j = \frac{1}{2} \sum_i W_{ij} - \theta_j. \tag{1.10}$$

They then each step towards their appropriate quadrant. There are several effects that will determine the final position of each word presentation. First, as is clear from our HCA and MDS plots, similar words will tend to follow similar trajectories and end up in similar regions of hidden unit space. High-frequency words of all types will tend to have had plenty of time to get well into the right quadrant. The positions for the lower-frequency words will be more variable. Words containing no ambiguity will head directly to the correct quadrants. Ambiguous phonemes in exception words and closely related regular words (often referred to as regular inconsistent words) will be pulled towards two (or more) different quadrants with strengths proportional to their relative frequencies. Although the network eventually learns to use the context information to resolve these ambiguities, these points will still be the last to cross into the right segments and hence be the ones left closest to the axes. Strange words (such as 'sieve'), that have very rare spelling patterns, may also be left near the axes depending on their word frequency. Such a pattern of learning is in broad agreement with that found in children [5]. Also, the pattern of performance that will arise if there are problems in completing the later stages of learning is consistent with developmental surface dyslexia in children [19].

These effects are seen clearly in figure 1.10. The points (circled) nearest the group borderlines tend to be exception words (e.g., 'pint'), regular inconsistents (e.g., 'hive') and homographs (e.g., 'wind'). The other points (arrowed) near the zero projection lines correspond to borderline cases in orthogonal directions (e.g., 'been' is a borderline case in the /i/–/E/ plane) or orthographically strange words (e.g., 'sieve'). Similarly, figure 1.11 shows the two-dimensional subspace corresponding to the /i/ and /f/ phonemes with the /i/, /I/ and /f/ words plotted.

Not surprisingly, given the virtual absence of ambiguity between the /i/ and /f/ phonemes, the groups are much further apart than the /i/ and /I/ phonemes were in figure 1.10.

We can also see how the /i/–/I/ distinction looks from orthogonal directions. It is often argued that there should be a correlation between network output activation error scores and the corresponding reaction times in humans [83]. This follows because, in the more realistic cascaded approach to modeling reaction times [10], the rate of output activation build-up is proportional to the appropriate projection. Thus, the closer each point falls to the axes of our projection graphs, the longer the corresponding reaction time. We can therefore easily read off from our graphs the model's predictions for naming latency experiments: there will be a basic frequency effect. High-frequency words will not show a type effect, low-frequency exception words will be slower than regular inconsistent words which will be slower than consistent regular words and strange words will also have an increased latency effect. These predictions do turn out to be in broad agreement with experiment.

The original reason for wanting to investigate our network's internal representations was to gain insight into how various forms of acquired dyslexia may occur in the model. Connectionist models that can deal with regular and exception words in a single system cast doubt on the traditional dual-route

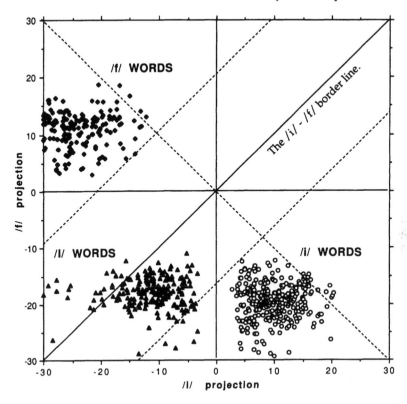

Figure 1.11. The 2D hidden unit sub-space corresponding to the /i/ and /f/ phonemes.

models of reading with their separate phonemic and lexical routes. However, a minimum requirement for them to replace the dual-route model completely is for them to be able to exhibit both surface dyslexia (lost exceptions) and phonological dyslexia (lost non-words) when damaged appropriately [19]. A detailed study of six representative forms of network damage [8] showed that, for each form of damage where there was a significant type effect (namely weight scaling, weight reduction, addition of noise to weights, removal of random connections and removal of random hidden units), we always find symptoms similar to surface dyslexia but never anything like phonological dyslexia. In each case we increased the degree of damage from zero to a level where the network failed to produce any correct outputs at all and patients with varying degrees of dyslexia corresponded to particular intermediate stages of this process.

How do we understand these results in terms of our projection plots? Since we are primarily concerned here with the internal representations we shall illustrate the analysis for just one form of damage, namely weight scaling. It is this form of damage in small networks that seems to give the most reliable

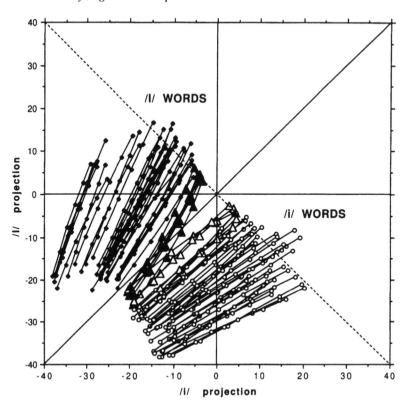

Figure 1.12. The effect of damage by weight scaling on our /i/–/I/ projections.

indication of what is likely to occur in more realistic networks [12]. We simply scale all the weights and thresholds by a constant scale factor $0 < \gamma < 1$. The effect of decreasing γ is to flatten all the sigmoids and, since the winning output phoneme is independent of the flatness of the output sigmoids, all the effect can be seen at the hidden units. As the hidden unit sigmoids are flattened, all the hidden unit activations tend to 0.5 and all the projections head back to the A_j defined above. It turns out that all the A_j are large and negative (mean -50.3, s.d. 10.2) so all the points drift more or less parallel to the bottom left diagonal. We see this clearly in figures 1.12 and 1.13.

The flow is fairly laminar, so the first points to cross the phoneme borders tend to be those that started off nearest to the borders. Thus the errors are predominantly on low-frequency exceptions rather than regular words and the errors for small amounts of damage tend to be regularizations. This is precisely the pattern of errors commonly found in surface dyslexics. Given this clear understanding of the effects of damage here we can be more confident in our claims about the effects of damage more generally [12].

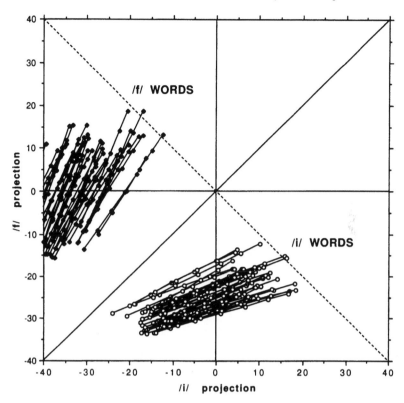

Figure 1.13. The effect of damage by weight scaling on our /i/–/f/ projections.

1.2.6   Contribution analysis

We have already mentioned briefly the important related questions of robustness and scaleability, i.e., are our networks sufficiently robust to damage and are our networks sufficiently large scale that we can confidently extrapolate their performance to larger networks? In this section we discuss contribution analysis, which is a technique for analyzing the importance of individual hidden units in connectionist networks [203].

In terms of our hidden unit activation $P_{i\alpha}$ for input pattern α, the standard network output activation can be written as in

$$\text{Out}_{j\alpha} = \text{Sigmoid}\left(\sum_i W_{ij} P_{i\alpha} - \theta_j\right). \tag{1.11}$$

Hence the contribution of hidden unit i to output unit j can be defined by

$$C_{ij\alpha} = \text{Sign}\left(\sum_i W_{ij} P_{i\alpha} - \theta_j\right) W_{ij} P_{i\alpha}. \tag{1.12}$$

We introduce the Sign function to ensure that positive contributions always enhance the accuracy of the output whether it be 0 or 1. Since the contributions $C_{ij\alpha}$ take into account the output weights, we should expect them to be more useful for analyzing the network's performance than the $P_{i\alpha}$ themselves. However, the extra index j means that we now have many times the number of components to deal with and these clearly have to be reduced for visualization purposes. Perhaps the most convenient and illuminating way to proceed is simply to perform analyses of the form described above on the contributions for particular output units or particular hidden units. The advantages of this approach were originally illustrated using PCA on a simplified reading model with one hidden layer [203]. More recently it has been shown how it can also be used to analyze multi-layer networks with cross-connections between hidden layers [82].

Another important use of contribution analysis, that we have not already discussed, is to examine the effect of single hidden units or connections on the output of the network. If individual contributions, or small numbers of contributions, have a significant effect on the network's outputs, then the network will not be robust with respect to damage and simulations of network damage will not necessarily scale up to more realistically sized networks. By definition, minimal networks will be highly dependent on individual contributions, whereas much larger networks performing the same task are likely to form more distributed internal representations with individual contributions insignificant with respect to the sum of the others.

This aspect of network analysis was discussed more fully [12] in the context of connectionist neuropsychology. The important variable here is the ratio C of each contribution compared with the total effect of all the contributions (including the threshold). If $C > 1.0$, then removing that contribution will change the sign of the total and (assuming the network has been trained to produce near-binary outputs) drastically change the output. Figure 1.14 shows (for a simple model) that the number of patterns affected by at least one such contribution does indeed decrease from 100% for minimal networks to zero for larger networks (e.g., with more than 150 hidden units in this case).

If $C > 1/N$ we run the risk that the removal of N random connections will have a significant effect on the outputs. Figure 1.14 also shows how the number of patterns affected by contributions with $C > 0.5$ and $C > 0.3$ falls with the number of hidden units. It is worrying that we need tens, if not hundreds, of times the minimal number of hidden units to provide a reasonably robust and distributed network.

1.3   Conclusions

We have surveyed a number of techniques for analyzing the internal representations (i.e., patterns of hidden unit activation) of trained feedforward networks and seen how they may all provide useful information concerning the

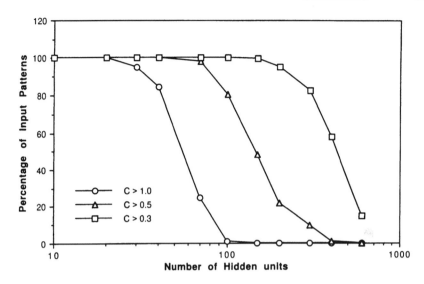

Figure 1.14. The number of patterns dependent on significant individual contributions.

operation of what would otherwise be 'black box' systems. In doing so we have also identified a number of pitfalls in these approaches which may result in misleading information about how the networks are performing.

We began by seeing how cluster analysis can indicate patterns of hidden unit activation that agree well with how we might expect the network to operate, but the mis-classification of exceptional items (that the network itself could handle correctly) highlighted the fact that this approach was missing out on some crucial aspects of the network's performance. We then turned to various techniques for reducing the dimension of the hidden unit activation space to something that could be visualized more directly. Firstly we used standard principal component analysis to reduce the number of dimensions with minimal loss of information, but found that for non-trivial problems two or three dimensions simply cannot capture enough of the variance for more than the grossest features to be recognized. We then saw how the non-metric approach of multi-dimensional scaling could provide a slightly better picture of what was happening. The problem with both these approaches is that neither takes account of the fact that some hidden units are more important than others simply because they are connected with different weights to the output layer. They consequently tend to mis-represent what is happening for some items, the exceptional items in particular. To remedy this problem we then looked for particular directions in the hidden unit activation space that corresponded to known features of the network's mapping. Using discriminant analysis we were able to find projection vectors that best discriminated between various classes of network outputs. However, careful analysis showed that this approach could indicate modes of network operation that did not actually correspond to how the network was really operating, and that this difficulty was particularly problematic when the network employed many more degrees of freedom than were necessary to perform the given task. It was then noted that, for networks where it was possible, simple projections onto sub-spaces defined by the output weights provide a much better picture of the network's internal representations than any of the preceding techniques. This is unfortunate, since output weight projections will not be nearly so simple for systems that have more complicated output representations. We ended with a brief discussion of contribution analysis and how it may be used with the preceding techniques and how it can be used to investigate the robustness and scaleability of trained networks.

Throughout this chapter we have used a simple reading model to illustrate the various techniques. We have seen explicitly in this case how an analysis of the internal representations can not only provide an insight into how the network is operating, but can also lead to a better understanding of various human developmental effects and reaction times. Similarly, analyzing the network's response to damage can lead to better models of acquired dyslexia. It seems likely that the simple techniques discussed in this chapter will be able to provide equally useful insights into the operation of a wide range of other connectionist systems.

Chapter 2

Information Maximization to Simplify Internal Representation

Ryotaro Kamimura
Tokai University, Japan
ryo@cc.u-tokai.ac.jp

2.1 Introduction

Many attempts have been made to reduce network size and to obtain suitable or optimal networks [135]. By reducing network size, obtained internal representations are simplified in terms of the number of input–hidden connections or hidden units. Thus it is much easier to interpret internal representations and the mechanism of learning, namely, how a network responds to input patterns. In addition, the reduction of network size consists essentially in the reduction of the number of parameters: generalization performance is expected to be improved.

Most of the methods proposed are concerned mainly with practical techniques simply to prune units or connections [66]. However, information maximization methods described in this chapter [50] are not only concerned with practical techniques but also with attempts to interpret the meaning of network size reduction and to view the network size reduction in more unified ways. In a framework of the information maximization methods, the network size reduction is considered to be a method aiming to amplify one aspect of neural learning, namely, information compression. By compressing information into several units or connections, it is possible to reduce the network size.

Information is defined as the decrease of uncertainty of hidden units for input patterns. By maximizing the information, internal representations can be simplified because of the information compression. Eventually, it is easier to interpret the representations. At this stage, we can easily decide which hidden unit should be pruned. Thus, network size reduction can be realized by condensing information into several hidden units in easily interpretable ways. The condensation is one of the most important aspects of information maximization methods.

2.2   Information maximization method

2.2.1   Outline of information maximization

Information maximization has been used to condense information into a few hidden units. By maximizing information, a few hidden units tend usually to be strongly activated, while all the other hidden units are inactive because of strongly negative input–hidden connections. Figure 2.1(a) depicts a process of information maximization in which only one hidden unit is forced to be turned on and the rest are off.

Information maximization methods are used to inhibit hidden units by strongly negative connections. Because of the negative connections, it happens that all the hidden units can approximately be off, as shown in figure 2.1(b). This means that if input patterns are not so important, all the hidden units tend to be off for the patterns. By this effect, the number of hidden units actually used tends to be reduced to a minimum number. Thus, by maximizing the information defined on hidden units, we obtain simplified networks in terms of the number of hidden units and, in some cases, rules behind input patterns can be extracted [50].

2.2.2   The concept of information

In this section, we briefly introduce information in a general framework of an information theory. Let Y take on a finite number of possible values $y_1, y_2, \ldots, y_M$ with probabilities $p(y_1), p(y_2), \ldots, p(y_M)$, respectively. Then, initial or prior uncertainty $H(Y)$ of a random variable Y is defined by

$$H(Y) = -\sum_{j=1}^{M} p(y_j) \log p(y_j). \tag{2.1}$$

Conditional uncertainty $H(Y \mid X)$ can be defined as

$$H(Y \mid X) = -\sum_{s=1}^{S} p(x_s) \sum_{j=1}^{M} p(y_j \mid x_s) \log p(y_j \mid x_s). \tag{2.2}$$

We can easily verify that conditional uncertainty is always less than or equal to initial uncertainty. Information is usually defined as the decrease of this uncertainty [4]:

$$I(Y \mid X) = H(Y) - H(Y \mid X)$$
$$= -\sum_{j=1}^{M} p(y_j) \log p(y_j) + \sum_{s=1}^{S} p(x_s) \sum_{j=1}^{M} p(y_j \mid x_s) \log p(y_j \mid x_s). \tag{2.3}$$

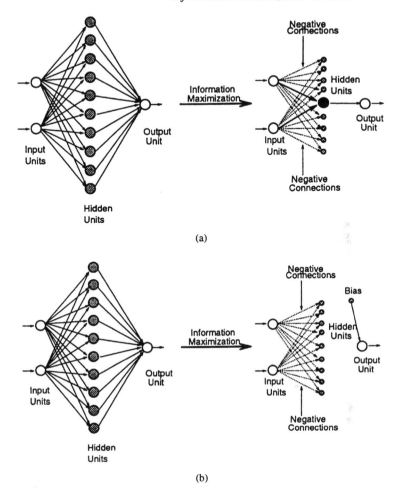

(a)

(b)

Figure 2.1. (a) Only one hidden unit is strongly activated by the information maximization; (b) all the hidden units are inactive because of strongly negative connections.

Especially, when a prior uncertainty is maximum, namely, a prior probability is equi-probable $(1/M)$, the information is:

$$I(A \mid B) = \log M + \sum_{s=1}^{S} p(x_s) \sum_{j=1}^{M} p(y_j \mid x_s) \log p(y_j \mid x_s) \qquad (2.4)$$

where $\log M$ is a maximum uncertainty. This form of information has been applied successfully to many problems because it is usually impossible to specify a prior probability before observation.

2.2.3   An information theoretic formulation

Let us define information defined in a general framework for an actual network shown in figure 2.2. Suppose that v_j^s, ranging between zero and one, is an output from the jth hidden unit, given the sth input pattern, and p_j^s is the jth normalized hidden unit activity defined by

$$p_j^s = \frac{v_j^s}{\sum_{m=1}^{M} v_m^s} \tag{2.5}$$

where M is the number of hidden units. Then, this normalized output p_j^s can be used to approximate a probability of the jth hidden unit, given the sth input pattern,

$$p(y_j \mid x_s) \approx p_j^s. \tag{2.6}$$

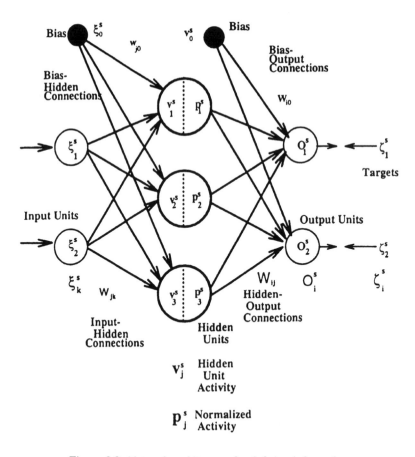

Figure 2.2. Network architecture for defining information.

At an initial state of the learning, hidden units respond uniformly to any input patterns. This means that the probability of hidden units is equi-probable, namely,

$$p(y_j) \approx \frac{1}{M}. \tag{2.7}$$

Input patterns are given to networks randomly, namely,

$$p(x_s) \approx \frac{1}{S}. \tag{2.8}$$

Thus, information can be computed by

$$
\begin{aligned}
I(Y \mid X) &= -\sum_{j=1}^{M} p(y_j) \log p(y_j) \\
&\quad + \sum_{s=1}^{S} \sum_{j=1}^{M} p(x_s) p(y_j \mid x_s) \log p(y_j \mid x_s) \\
&\approx -\sum_{j=1}^{M} \frac{1}{M} \log \frac{1}{M} + \sum_{s=1}^{S} \frac{1}{S} \sum_{j=1}^{M} p_j^s \log p_j^s \\
&= \log M + \frac{1}{S} \sum_{s=1}^{S} \sum_{j=1}^{M} p_j^s \log p_j^s
\end{aligned}
\tag{2.9}
$$

where $\log M$ is a maximum entropy.

By changing the information, different internal representations with different information can be obtained. When the information is maximized, only one hidden unit is strongly activated for a given input pattern, while all the other hidden units are inactive. Our experimental results confirm that if all the hidden units are not very important, all the hidden units tend to be inactive by maximizing the information. As shown in figure 2.1(b), all the hidden units are approximately turned off in a maximum information state. This is easily explained by seeing a normalized activity p_j^s. If a hidden unit is *relatively* larger than the other hidden units, the information can be maximized. By using this property of the information function, the information maximization can turn off unnecessary hidden units and turn on necessary hidden units.

2.2.4 Information maximization

Since we have already defined an information function, it is necessary to formulate update rules to increase this information. Now, suppose that a network is composed of three layers: input, hidden and output layers as in figure 2.2. An output from the jth hidden unit, given the sth input pattern, is denoted by v_j^s. The kth element of the sth input pattern is given by ξ_k^s. A connection from the kth input unit to the jth hidden unit is denoted by w_{jk}. A connection from the jth hidden unit to the ith output unit is denoted by W_{ij}. The jth hidden unit produces an output

$$v_j^s = f(u_j^s) \tag{2.10}$$

where u_j^s is a net input into the jth hidden unit and is computed by

$$u_j^s = \sum_{k=0}^{L} w_{jk} \xi_k^s \tag{2.11}$$

where ξ_k^s is the kth element of the sth input pattern, L is the number of elements in the pattern and f is a sigmoid activation function defined by

$$f(u_j^s) = \frac{1}{1 + \exp(-u_j^s)}. \tag{2.12}$$

Thus, information has been computed by

$$I = \log M + \frac{1}{S} \sum_{s=1}^{S} \left(\sum_{j=1}^{M} p_j^s \log p_j^s \right). \tag{2.13}$$

Differentiating the information with respect to input–hidden connections, we have,

$$
\begin{aligned}
S \frac{\partial I}{\partial w_{jk}} &= \sum_{s=1}^{S} \sum_{m=1}^{M} \frac{\partial I}{\partial p_m^s} \frac{\partial p_m^s}{\partial v_j^s} \frac{\partial v_j^s}{\partial w_{jk}} \\
&= \sum_{s=1}^{S} \sum_{m=1}^{M} \left\{ (1 + \log p_m^s) \frac{\delta_{jm} \sum_t v_t^s - v_m^s}{(\sum_t v_t^s)^2} \right\} f'(u_j^s) \xi_k^s \\
&= \sum_{s=1}^{S} \left(\log p_j^s - \sum_{m=1}^{M} p_m^s \log p_m^s \right) p_j^s (1 - v_j^s) \xi_k^s \\
&= \sum_s \phi_j^s \xi_k^s
\end{aligned}
\tag{2.14}
$$

where

$$\phi_j^s = \left(\log p_j^s - \sum_{m=1}^{M} p_m^s \log p_m^s \right) p_j^s (1 - v_j^s). \tag{2.15}$$

Update rules just formulated can be used to control the information, especially to maximize the information. For illustration, the number of hidden units is supposed to be two. Then, we define a Φ-function as follows:

$$\phi_j^s = \Phi_j^s (1 - v_j^s) \tag{2.16}$$

where

$$\Phi_j^s = \left\{ \log p_j^s - p_j^s \log p_j^s - (1 - p_j^s) \log(1 - p_j^s) \right\} p_j^s. \tag{2.17}$$

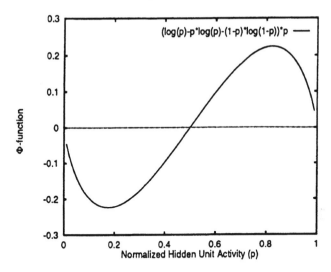

Figure 2.3. A function Φ_j^s as a function of a normalized hidden unit output p_j^s.

Figure 2.3 shows Φ_j^s as a function of p_j^s. If a normalized activity is over 0.5, connections are pushed toward positive connections, to turn a unit on. If the activity is less than 0.5, the connections are pushed toward negative connections, forcing the hidden unit to be turned off.

Information is maximized under the condition that errors between targets and outputs are sufficiently small. Thus, in addition to the ϕ function, the δ rules for decreasing the errors must be incorporated when input–hidden connections are updated. Rules for updating are

$$\Delta w_{jk} = \beta S \frac{\partial I}{\partial w_{jk}} - \eta S \frac{\partial D}{\partial w_{jk}}$$

$$= \beta \sum_{s=1}^{S} \phi_j^s \, \xi_k^s + \eta \sum_{s=1}^{S} \delta_j^s \, \xi_k^s \qquad (2.18)$$

where β and η are parameters, D is a cross entropy cost function and δ_j^s is a delta function for a cross entropy function defined in the next section.

2.2.5 Cross entropy minimization

For the definition of cross entropy, we need conditional probabilities, given input patterns. One of the candidates for the probability is an output from the ith output unit computed by

$$O_i^s = f \left(\sum_{j=0}^{M} W_{ij} v_j^s \right). \qquad (2.19)$$

This actual output is used to approximate the probability of the firing of the ith output unit, given the sth input pattern. A desired probability is approximated by a target ζ_i^s for the corresponding output. In this formulation, all the output units are considered to be independent of each other. A probability of occurrence of input patterns is considered to be equi-probable, namely, $1/S$. Thus, cross entropy is computed by

$$D = \frac{1}{S} \sum_{i=1}^{N} \left[\sum_{s=1}^{S} \left\{ \zeta_i^s \log \frac{\zeta_i^s}{O_i^s} + (1 - \zeta_i^s) \log \frac{1 - \zeta_i^s}{1 - O_i^s} \right\} \right] \tag{2.20}$$

where ζ_i^s is a target for the output O_i^s from the ith output unit, N is the number of output units and input units and S is the number of input patterns [91]. If the difference between targets and outputs is smaller, the cross entropy function is decreased. For hidden–output connections, we have rules for updating

$$\Delta W_{ij} = -\eta S \frac{\partial D}{\partial W_{ij}}$$

$$= \eta \sum_{s=1}^{S} (\zeta_i^s - O_i^s) v_j^s$$

$$= \eta \sum_{s=1}^{S} \delta_i^s v_j^s \tag{2.21}$$

where δ_i^s is computed by

$$\delta_i^s = \zeta_i^s - O_i^s. \tag{2.22}$$

For input–hidden connections, we have

$$\Delta w_{jk} = -\eta S \frac{\partial G}{\partial w_{jk}}$$

$$= \eta \sum_{s=1}^{S} v_j^s (1 - v_j^s) \sum_{i=1}^{N} W_{ij} \delta_i^s \xi_k^s$$

$$= \eta \sum_{s=1}^{S} \delta_j^s \xi_k^s \tag{2.23}$$

where

$$\delta_j^s = v_j^s (1 - v_j^s) \sum_{i=1}^{N} W_{ij} \delta_i^s. \tag{2.24}$$

2.3 Modified information maximization methods

2.3.1 Outline of modified methods

One of the shortcomings of information maximization methods formulated in the previous section is that as information for hidden units is increased, hidden units tend to be over-saturated, meaning that outputs from hidden units are close to one, and the magnitude of input–hidden connections is overwhelmingly large. There is plenty of room for over-fitting by adjusting the magnitude of each input–hidden connection. For overcoming this, we propose several modified maximum information methods for improving generalization performance and for simplifying internal representations. The main strategy of the modified methods consists in introducing an information maximizer, a specialized unit for maximizing information. The information maximizer is used only to maximize information. All the other input–hidden connections are not used to maximize the information. By this modification, the excessive generation of negative connections is prohibited, because only connections from the information maximizers become strongly negative. Thus, the information maximization is used to improve generalization performance as well as to interpret internal representations.

2.3.2 Information maximizer

To prohibit the generation of multiple negative connections, we have introduced an information maximizer whose function consists exclusively in maximizing the information. Let d_j be a connection from an information maximizer to the jth hidden unit: a net input into the jth hidden unit is computed by

$$u_j^s = \sum_{k=0}^{L} w_{jk}\xi_k^s + d_j. \tag{2.25}$$

As shown in figure 2.4(a), information is maximized only by updating these connections from the information maximizer. Rules for updating are obtained by differentiating the information function with respect to connections from the information maximizer

$$\Delta d_j = \beta S\frac{\partial I}{\partial d_j} = \beta \sum_{s=1}^{S} \phi_j^s. \tag{2.26}$$

The information maximizer and the bias can be unified as shown in figure 2.4(b). Connections from the information maximizer are used to maximize information and to minimize the cross entropy function. A net input into the jth hidden unit is computed by

$$u_j^s = \sum_{k=1}^{L} w_{jk}\xi_k^s + d_j. \tag{2.27}$$

and rules for updating are formulated by

$$\Delta d_j = \beta S\frac{\partial I}{\partial d_j} - \eta S\frac{\partial G}{\partial d_j} = \beta \sum_{s=1}^{S} \phi_j^s + \eta \sum_{s=1}^{S} \delta_j^s. \tag{2.28}$$

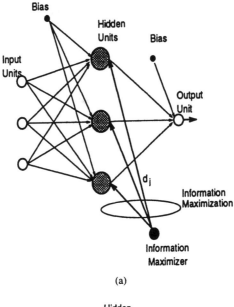

(a)

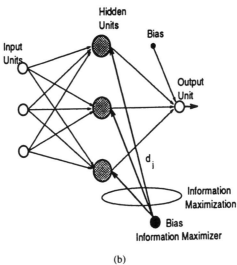

(b)

Figure 2.4. (a) Information maximizer and bias are separated; (b) information maximizer and bias are unified.

For all the other input–hidden connections, update rules are

$$\Delta w_{jk} = \eta \sum_{s=1}^{S} \delta_j^s \xi_k^s. \qquad (2.29)$$

2.3.3 Information maximizer with weight elimination

We have presented a method to maximize information by using an information maximizer. Since the information maximizer operates independently of the other input–hidden connections, weight decay or a weight elimination term can be incorporated to reduce the number of input–hidden connections.

Let us suppose that an information maximizer is combined with bias, as shown in figure 2.5(b). The weight decay term is defined by

$$T = \frac{1}{2} \sum_{j=1}^{M} \sum_{k=1}^{L} w_{jk}^2 \tag{2.30}$$

where the summation is over all the input units and hidden units. Another penalty term is the weight elimination penalty term [102]. It is defined by

$$Q = \frac{1}{2} \sum_{j=1}^{M} \sum_{k=1}^{L} \frac{w_{jk}^2}{1 + w_{jk}^2}. \tag{2.31}$$

Thus, rules for updating for input–hidden connections by the weight decay are formulated by

$$\Delta w_{jk} = -\eta S \frac{\partial D}{\partial w_{jk}} - \gamma \frac{\partial T}{\partial w_{jk}}$$

$$= \eta \sum_{s=1}^{S} \delta_j^s \xi_k^s - \gamma w_{jk} \tag{2.32}$$

where γ is a learning parameter. Rules for the weight elimination are defined by

$$\Delta w_{jk} = -\eta S \frac{\partial D}{\partial w_{jk}} - \gamma \frac{\partial Q}{\partial w_{jk}}$$

$$= \eta \sum_{s=1}^{S} \delta_j^s \xi_k^s - \gamma \frac{w_{jk}}{(1 + w_{jk}^2)^2}. \tag{2.33}$$

On the other hand, for connections from an information maximizer, information is maximized by the following rules:

$$\Delta d_j = \beta S \frac{\partial I}{\partial d_j} - \eta S \frac{\partial G}{\partial d_j} = \beta \sum_{s=1}^{S} \phi_j^s + \eta \sum_{s=1}^{S} \delta_j^s. \tag{2.34}$$

We have presented a method in which an information maximizer and bias are unified. Note that the maximizer and bias can separately be used as shown in figure 2.5(a).

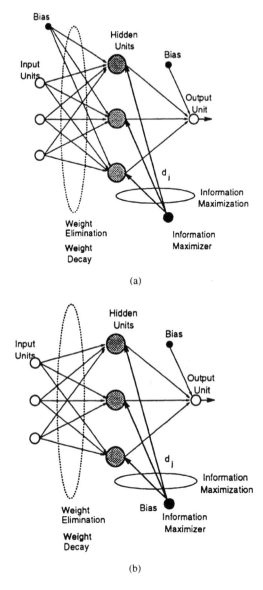

Figure 2.5. Information maximization with weight decay and weight elimination: (a) information maximizer and bias are separated; (b) information maximizer and bias are unified.

2.4   Application to the XOR problem

The information maximization method was applied to the familiar XOR problem. In this problem, we examined whether the information maximization method

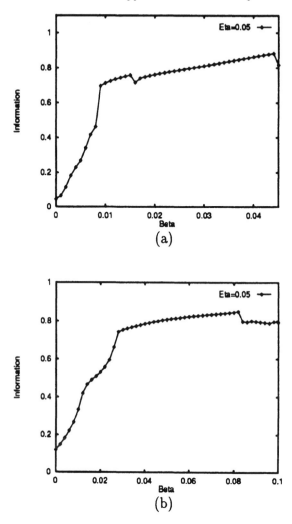

Figure 2.6. Information as a function of a parameter β for the XOR problem: (a) a standard information maximization method; (b) information maximizer.

could produce a minimal network. In experiments, the number of input, hidden and output units was two, ten and one unit. A learning rate η was 0.05 for all the experiments. The learning was considered to be finished when the absolute difference between targets and outputs was less than 0.1 for all the input patterns.

Figure 2.6 shows information as a function of a parameter β. Relative information or normalized information was computed by

$$I^{rel} = \frac{I}{I^{max}} = \frac{I}{\log M} \tag{2.35}$$

where $\log M$ is a maximum entropy. Thus, the relative information ranges between zero and one. Figure 2.6(a) shows relative information by a standard information maximization method. Information was 0.03 without the information maximization method. Information is increased as a parameter β is increased and a maximum point was 0.88 close to maximum information, when parameter β was 0.044. Then, a modified information maximization method is used in which an information maximizer and bias are separately used. Figure 2.6(b) shows information obtained by a modified information maximization method. Information is increased up to 0.85 for parameter $\beta = 0.082$. Thus, maximum information by the modified information maximization is slightly lower than maximum information by a standard information maximization method.

Table 2.1.   Relevance and error for the XOR problem obtained by information maximization.

Hidden Unit	Relevance	Error
1	1.4	0.3
2	0.6	0.0
3	0.0	0.0
4	0.0	0.0
5	0.0	0.0

Table 2.1 shows a measure of relevance and an error rate for the XOR problem obtained by a standard information maximization method. Before going into detail, we need some explanation for the measure [66]. The relevance is used to measure how important a given hidden unit is for producing appropriate outputs correctly. The relevance (R_m) of the mth hidden unit is defined by

$$R_m = \sum_{s=1}^{S} \sum_{i=1}^{N} \left\{ \zeta_i^s - f \left(\sum_{j=0}^{M} W_{ij} \rho_{mj} v_j^s \right) \right\}^2 - \sum_{s=1}^{S} \sum_{i=1}^{N} (\zeta_i^s - O_i^s)^2 \qquad (2.36)$$

where ρ_{mj} is 0 for $m = j$ and one, otherwise. Thus, the relevance of the mth hidden unit is an error computed without the mth hidden unit. If the relevance is large, the hidden unit plays a very important role. Figure 2.7(a) shows a case in which the error between targets and outputs without the first hidden unit is significantly large, and it is considered to be an important hidden unit. On the other hand, figure 2.7(b) shows an unimportant hidden unit without which a network produces only small errors between targets and outputs.

Now, let us turn to the explanation of table 2.1. In the table, hidden units are ordered according to the magnitude of the relevance. The error rate is an error computed with the corresponding number of hidden units. We considered outputs from output units to be one, when they are greater than 0.6. On the

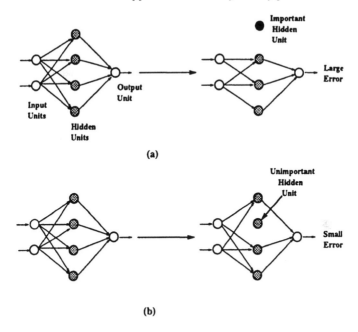

Figure 2.7. Detection of (a) an important hidden unit and (b) an unimportant hidden unit.

other hand, outputs are considered to be zero, when they are less than 0.4. An error rate means how many input patterns can unsuccessfully be estimated by neural networks. For example, an error rate of 0.25 means that only one input pattern out of four input patterns cannot be successfully estimated. In the table, the error is 0.3 only with the first hidden unit with the largest relevance. The error becomes zero with the first and the second hidden units. Thus, only two hidden units are needed to produce appropriate outputs.

Figure 2.8 shows a network obtained by an information maximization method. In figure 2.8(a), we can see that almost all the input–hidden connections are negative connections represented by dotted lines. They tend to make almost all the hidden units inactive. Only two units are detected to be important. Figure 2.8(b) explicitly shows that the number of hidden units can be reduced to two, a minimum number of hidden units.

2.5 Application to the symmetry problem

The second problem for experiments is the *symmetry problem* in eight bits [81]. It is well known that only two hidden units are needed to learn this symmetry problem, and in addition typical symmetric connections are generated as input– hidden connections. For experiments, eight input units, eight hidden units and

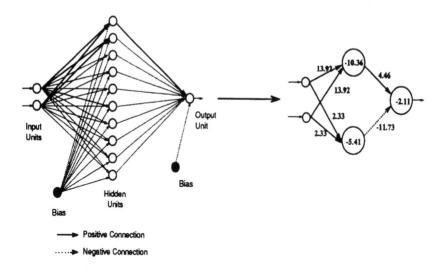

Figure 2.8. Network size reduction by information maximization for the XOR problem.

one output unit were used. The learning rate η was 0.05 for all the experiments. The learning was considered to be finished when the absolute difference between targets and outputs was less than 0.2 for all the input patterns.

Table 2.2. Relevance and error for the symmetry problem, obtained with standard information maximization.

Hidden unit	Relevance	Error
1	93.3	0.5
2	93.2	0.0
3	0.0	0.0
4	0.0	0.0
5	0.0	0.0

First, information was increased as much as possible. Figure 2.9(a) shows relative information as a function of a parameter β for a set of initial conditions by a standard information maximization method. The information is gradually increased, as the parameter β is increased. A highest point of the information was 0.94, when the parameter β was 6×10^{-5} for a set of initial values. This state was significantly close to a final state of maximum information. Then, a modified information maximization method was used in which an information maximizer and bias are completely separated. Though information is significantly increased, obtained maximum information was 0.93

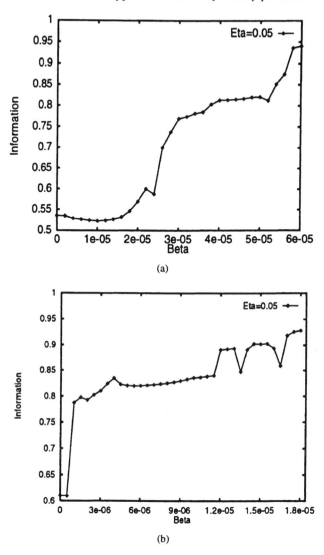

Figure 2.9. Relative information as a function of a parameter β for the symmetry problem: (a) standard information maximization method; (b) modified information maximization method.

($\beta = 1.8 \times 10^{-5}$), which is slightly lower than maximum information obtained by a standard information maximization method.

Table 2.2 shows relevance for five hidden units of ten hidden units. The relevance of all the remaining five hidden units was close to zero. Hidden units are arranged in decreasing order of relevance. The first two hidden units have the

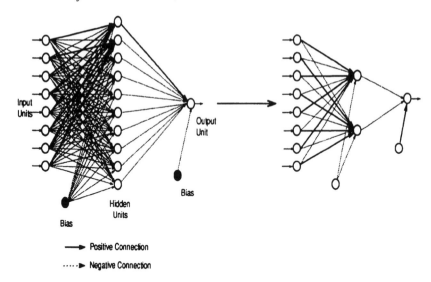

Input
Units

Output
Unit

Bias

Hidden
Units

Bias

——▶ Positive Connection

·····▶ Negative Connection

Figure 2.10. Transformation of an original network into a simple network: (a) by maximizing information multiple negative connections are generated, making many hidden units inactive for all the input patterns; (b) eventually a simple network is generated where symmetric connections can be seen.

largest relevance, namely, 93.3 and 93.2. Actually, only these two hidden units were necessary for producing correct targets, because the error rate is completely zero with these hidden units, as shown in the table.

Then, let us show how the information maximization transforms an original oversized network. Figure 2.10 shows a network obtained by a standard information maximization, when information is close to a maximum. In figure 2.10(a), it can be seen that many strongly inhibitory connections are generated by the information maximization. By these inhibitory connections, many hidden units are eventually inactive for all the input patterns, which are eliminated without affecting the performance. In figure 2.10(b), only two hidden units can be detected. With these two hidden units, we can construct a simple network. Here symmetric connections [81] can clearly be seen.

2.6 Application to well-formedness estimation

2.6.1 Well-formedness by sonority

In this section, we present experimental results on the estimation of well-formedness of consonant clusters at the beginning of words according to a sonority principle. The sonority principle states that each consonant has its own sonority value, as shown in table 2.3 [57, 68]. The principle can be summarized as follows. Suppose that two consonants C_1, C_2 must be placed at the beginning

Table 2.3. Sonority hierarchy of an artificial language.

Order	Features	Examples
1	Voiceless stop	[p, t, k]
2	Voiced stop	[b, d, g]
3	Voiceless fricative	[f, θ, s]
4	Voiced fricative	[v, δ, z]
5	Nasal	[m, n]
6	Lateral	[l]
7	Retroflex	[r]
8	Semivowel	[y, w]

of words. If the sonority of a precedent consonant C_1 is smaller than that of the following consonant C_2, then the consonant cluster is well formed or permitted at the beginning of words. Otherwise, the consonant cluster is not permitted at the beginning of words. It is ill formed. For example, let us take an English word *predict*. In this example, a consonant cluster /pr/ at the beginning of words is well formed, because the sonority of /p/ (sonority = 1) is less than the sonority of /r/ (sonority = 7). On the other hand, a consonant cluster /rp/ is ill formed, because the sonority of /p/ is less than the sonority of /r/. In our experiments, each phoneme is represented in a phonological representation as shown in table 2.4. Figure 2.11 shows two examples of the well-formedness inference. Figure 2.11(a) shows an example of well-formedness in which a phoneme string /pr/ is well formed and an output is *Yes*. Figure 2.11(b) shows an example of ill-formedness in which a phoneme string /rp/ is given to a network and the network must produce *No*. As shown in figure 2.11, ten input, ten hidden and one output unit were used in the experiments.

2.6.2 Information and generalization

In experiments, we firstly examine whether information can be increased. Then, generalization performance is examined for 100 training and 100 testing patterns. Figure 2.12(a) shows relative information as a function of the parameter β. Here information is increased gradually up to 0.93 ($\beta = 0.0006$) close to a maximum point. Figure 2.12(b) shows generalization errors as a function of a parameter β. Generalization errors are increased in direct proportion to the increase of information. This is because by maximizing information, strongly negative input–hidden connections are generated, producing the saturation of hidden units and degrading generalization performance. If saturation by the negative connections is reduced, better generalization can be expected, though the effect of the information maximization term may be weakened as already discussed in the previous sections.

Table 2.4. Phonological representation used in our experiments [213].

Sonority	Phoneme	Voicing	Manner	Place
1	/p/	0	1 1	1 1
1	/t/	0	1 1	1 0
1	/k/	0	1 1	0 0
2	/b/	1	1 1	1 1
2	/d/	1	1 1	1 0
2	/g/	1	1 1	0 0
3	/f/	0	1 0	1 1
3	/θ/	0	1 0	1 0
3	/s/	0	1 0	0 1
4	/v/	1	1 0	1 1
4	/δ/	1	1 0	1 0
4	/z/	1	1 0	0 1
5	/m/	1	0 0	1 1
5	/n/	1	0 0	1 0
5	/η/	1	0 0	0 0
6	/l/	1	0 1	1 0
7	/r/	1	0 1	0 1
8	/y/	1	0 1	0 0
8	/w/	1	0 1	1 1

To prohibit the generation of multiple negative connections, we firstly used modified information maximization methods in which an information maximizer is introduced. In experiments, the information maximizer and bias are unified. Figure 2.13 shows information and generalization errors as a function of a parameter β by a modified information maximization method. Figure 2.13(a) shows relative information. The information is increased up to 0.90 for a parameter $\gamma = 0.0016$ and remains unchanged beyond that point. Figure 2.13(b) shows generalization errors as a function of a parameter β. Generalization errors are initially increased from 0.096 ($\beta = 0$) to 0.126 ($\beta = 0.009$). Then, the errors are decreased and remain unchanged. This means that generalization performance is not degraded even if information is maximized.

2.6.3 Interpretation of internal representation

Table 2.5 shows the relevance and the error for the well-formedness estimation. Only the first two hidden units have the highest relevance values and the relevances of all the other hidden units are close to zero. Since the magnitude (40.0) of the first hidden unit is much larger than the magnitude (5.8) of the second hidden unit, the first hidden unit performs a principal role in estimating the well-formedness of consonant clusters. In the table, we can see that the error

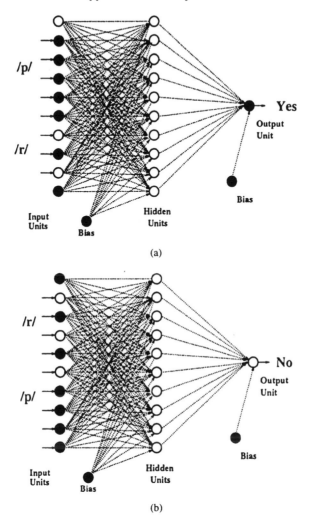

Figure 2.11. Examples of well-formedness inference: (a) a string /pr/ is well formed and the network should be trained to produce *Yes*; (b) a string /rp/ is ill formed and the network should be trained to produce *No*.

rate is only 0.1, which means that ninety per cent of 100 training consonant clusters can be explained by the first hidden unit.

Figure 2.14 shows an internal representation obtained by the information maximization in which hidden units are ordered from the top to the bottom according to the magnitude of the relevance. Here only some input–hidden connections into the first and the second hidden unit can take positive values. On the other hand, connections from an information maximizer are all strongly

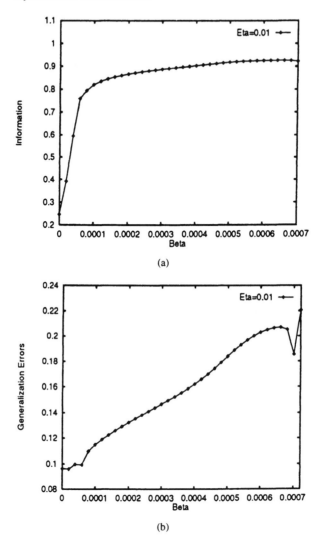

Figure 2.12. Relative information and generalization errors by standard information maximization for the consonant cluster estimation: (a) shows relative information as a function of a parameter β; (b) shows generalization errors in RMS as a function of a parameter β.

negative connections represented by dotted lines. Thus, the information maximizer is a generator of negative connections, prohibiting many hidden units from being activated.

Figures 2.15(a) and (b) show all the connections into and from the first hidden units. A hidden–output connection is a negative connection (-4.67) and

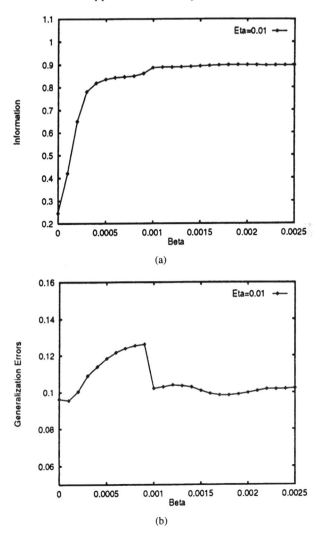

Figure 2.13. Relative information and generalization errors by an information maximizer: (a) relative information as a function of the parameter β; (b) generalization errors in RMS as a function of the parameter β.

bias to an output unit is positive (2.36). Thus, for well-formed clusters, the hidden unit should be off and for ill-formed clusters, the hidden unit should be strongly activated.

Let us interpret a mechanism of estimation of the well-formedness problem. First, the first input unit and the sixth input unit represent a *voicing* feature for the first and the second consonant. A connection from the first input

Table 2.5. Relevance and error for well-formedness inference obtained by information maximization.

Hidden unit	Relevance	Error
1	40.0	0.1
2	5.8	0.0
3	0.0	0.0
4	0.0	0.0
5	0.0	0.0

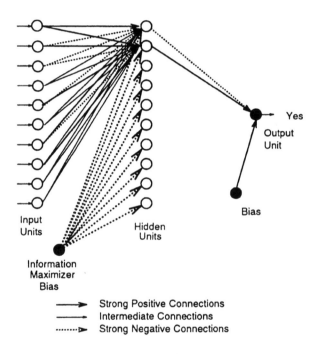

Figure 2.14. Internal representation obtained from information maximization.

unit is strongly positive (4.05) and a connection from the sixth input unit is strongly negative (−3.90). This means that if the first consonant is a *voiced* consonant and the second consonant is a *voiceless* consonant, the hidden unit has a high possibility to be strongly activated, meaning that the consonant cluster is ill formed. As shown in table 2.4, *voiced* consonants tend to have high sonority values. Secondly, connections from the second and third input units are strongly negative (−9.98 and −6.55), while connections from the seventh and the eighth input unit are strongly positive (3.49 and 6.59). This means that if *stop* consonants are the first consonants in consonant clusters, then the clusters are well formed. On the other hand, if *stop* consonants are the second consonants in consonant clusters, the clusters tend to be ill formed. As shown in table 2.3, *stop* consonants are supposed to have the lowest sonority values.

For example, in figure 2.15(a), a consonant cluster /pr/ is given to a network. Since the first consonant is a *voiceless* and *stop* consonant, and the second consonant is a *voiced* and *retroflex* consonant, then a hidden unit is inactive by strong negative connections and an output unit is strongly active by the bias to the output unit, meaning that the consonant cluster is well formed. In figure 2.15(b), a consonant cluster /rp/ is given to the network. Since the second consonant is a *stop* consonant, by the strongly positive connections and the strongly positive bias to the hidden unit, the hidden unit is on and the output unit is off, meaning that the consonant cluster is ill formed.

The second hidden unit deals with cases where the first hidden unit is on, even if a given consonant cluster is well formed. The second hidden unit is especially used to deal with consonants with similar sonority values. Figure 2.16 shows a case in which two consonants /n/ and /l/ have similar sonority values, namely, five and six, as shown in table 2.3. From our sonority principle, this cluster is well formed, because the sonority of the consonant /n/ is less than the sonority of the consonant /l/. However, the hidden unit is on and the output unit is off, namely, ill formed. On the other hand, the second hidden unit is strongly activated and a hidden–output connection (4.92) is larger than a hidden–output connection (-4.67) from the first hidden unit. Then, the output unit is strongly activated by the bias to the output (2.36). Thus, the second hidden unit helps networks to discriminate between consonants with similar sonority values.

2.6.4 Improved generalization with weight elimination

In this section, we use another method to maximize information and to improve generalization performance. This method consists mainly in introducing a weight decay term and a weight elimination term for updating input–hidden connections. In the modified information maximization method, information is maximized only by updating connections from an information maximizer. Thus, it is expected that the introduction of weight decay and weight elimination terms does not affect the process of information maximization. We use a modified information maximization method in which an information maximizer and the bias are unified, as described in figure 2.5(b).

Figure 2.17(a) shows relative information as a function of a parameter γ by introducing a weight decay term. The parameter β for the information term was fixed to be 0.0005. Here information tends to be increased in spite of the increased value of parameter γ. Maximum information was 0.89 for a parameter 1.4×10^{-6}. Figure 2.17(b) shows generalization errors as a function of a parameter γ. Generalization errors are clearly decreased as a parameter is larger for small values of the parameter γ. The minimum generalization

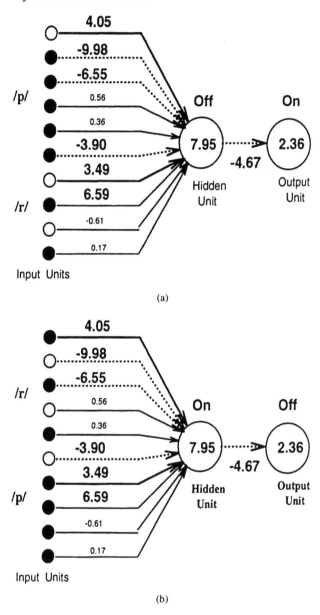

Figure 2.15. Estimation of well-formedness of the first hidden unit: (a) well-formedness; (b) ill-formedness.

error was 0.091 for a parameter 8×10^{-7}. Since the generalization error is 0.114 without the weight decay term, generalization performance is significantly improved.

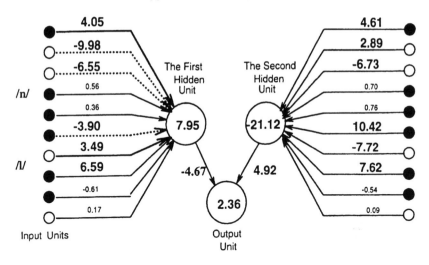

Figure 2.16. Estimation of well-formedness by the first and second hidden units; the latter deals with the detection of well-formedness of consonants with similar sonority values.

Then, for the same parameter value, a weight elimination penalty term was introduced. Figure 2.18(a) shows information as a function of the parameter γ. Here information is increased as parameter γ is increased. Maximum information was 0.86 for the parameter 2×10^{-5}. In addition, it could be seen that generalization performance is significantly improved. Figure 2.18(b) shows generalization errors. The generalization errors are rapidly decreased and close to 0.086 for the parameter 2×10^{-5}. Since the best generalization error is 0.09 by the weight decay, generalization performance is much improved by the introduction of the weight elimination term.

Then, parameter β was increased from 0.0005 to 0.001. Figure 2.19(a) shows relative information as a function of parameter γ by introducing a weight decay term. Here information is increased at an initial stage and reaches a maximum point of 0.89 for a parameter $\gamma = 4 \times 10^{-7}$. Then, information is decreased rapidly as parameter γ is increased. Figure 2.19(b) shows generalization errors as a function of parameter γ. Generalization errors are decreased at an initial stage and reach a minimum point of 0.089 for $\gamma = 6 \times 10^{-7}$. Then, generalization is inversely increased in direct proportion to information.

Figure 2.20(a) shows information by a weight elimination term. Here information is significantly increased up to 0.92 for a parameter $\beta = 2 \times 10^{-5}$, which is much higher than maximum information obtained simply by the information maximization. Figure 2.20(b) shows generalization errors as a function of parameter γ. Generalization errors are significantly decreased and

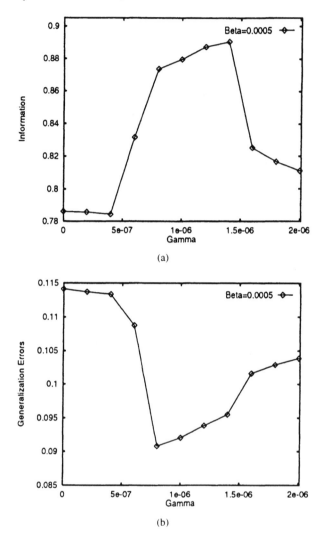

Figure 2.17. (a) Relative information; (b) generalization errors in RMS as a function of parameter γ for a well-formedness problem by an information maximizer with a weight decay method. The parameter β was 0.005.

finally close to 0.086 for the parameter $\gamma = 2 \times 10^{-5}$. These results show that by using the weight decay and the weight elimination term, information is not decreased and generalization performance is significantly improved.

Let us examine an internal representation obtained by a weight elimination term. Table 2.6 shows the relevance and the error by the information maximization with a weight elimination method. Though the relevance of the

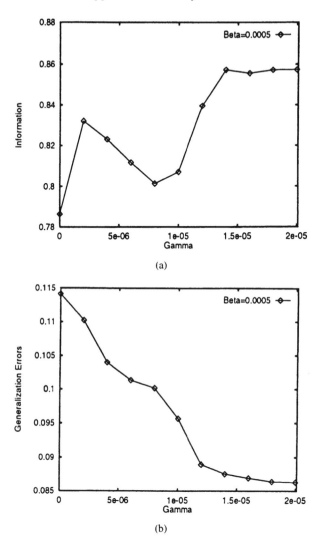

Figure 2.18. (a) Relative information; (b) generalization errors as a function of parameter γ for a well-formedness problem by a weight elimination term. The parameter β was 0.0005.

second hidden unit is slightly weakened and the relevance of the third hidden unit is increased from zero to 0.1, only two hidden units are needed to produce appropriate outputs.

Figure 2.21 shows an internal representation obtained by the information maximizer with the weight elimination. Values of connections and bias are approximately equivalent to those obtained by a standard information

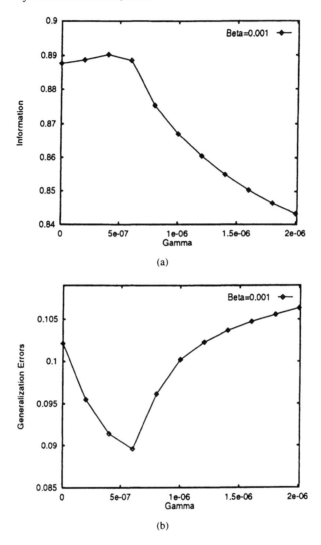

Figure 2.19. (a) Relative information; (b) generalization errors as a function of parameter γ for a well-formedness problem by incorporating a weight decay term. The parameter β was 0.001.

maximization method shown in figures 2.15 and 2.16. However, it can be seen that unnecessary input–hidden connections are reduced much further to smaller values. For example, input–hidden connections from the fourth and fifth input units are 0.13 and -0.18 respectively, while they are 0.56 and 0.36 without the weight elimination term.

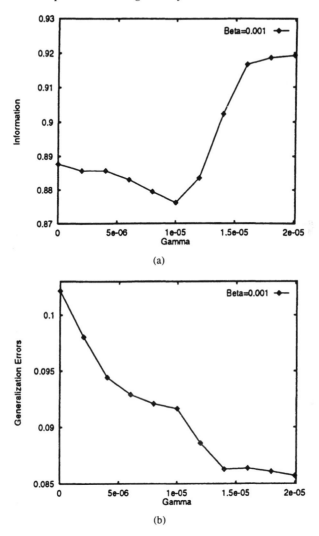

Figure 2.20. (a) Relative information; (b) generalization errors as a function of the parameter γ for the well-formedness problem by the weight elimination method. The parameter β was 0.001.

2.7   Competitive learning and information maximization

We have presented experimental results on the application of the information maximization method. Experimental results suggest that many aspects of neural learning in addition to the network size reduction can be described by the information maximization method. Let us take an example of the other possible applications of the information maximization method. Competitive learning

Table 2.6. Relevance and error for well-formedness inference obtained by information maximization with weight elimination.

Hidden unit	Relevance	Error
1	40.0	0.1
2	5.7	0.0
3	0.1	0.0
4	0.0	0.0
5	0.0	0.0

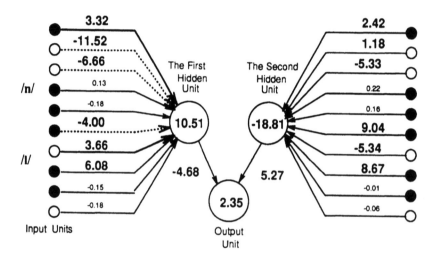

Figure 2.21. Estimation of the well-formedness by the first and second hidden units obtained by information maximization with weight elimination.

is a typical example of the possible applications. The competitive learning approach has been used as a feature detector or pattern classifier. One of the simplest forms was developed by Grossberg [41] and Rumelhart and Zipster [214]. In several other learning algorithms, for example Kohonen's feature map [52], ART [14], and counter propagation networks [69], competitive learning is incorporated as a substructure. Competitive learning is used to parse input patterns into distinct categories through the adaptive tuning of weights [41]. In a competitive layer, the unit with the largest weighted sum wins the competition, and is turned on as a winner, while all the other units in the competitive layer are turned off as losers (*winner-take-all*). Then, weights to winning units can be updated and weights become gradually similar to the corresponding input patterns in learning. The competitive learning method certainly shows a process of information maximization in the competitive layer. Given an appropriate formulation of an information function for a competitive layer, at a state of maximum information in the competitive layer, only one unit is turned on, while all the other units are off. On the other hand, if information is minimized, all the units in the competitive layer are equally activated. Thus, in the competitive learning paradigm, the information of the competitive layer must be maximized under a condition that weights must be similar to the corresponding input values in learning. The main point of information maximization is that an information function of a given layer must be maximized under certain conditions. Thus, the ordinary competitive learning is considered to be one of the information maximization methods in which information is maximized while weights into winning competitive units are similar to the corresponding input patterns. Thus, our information maximization method is a more general and more physiologically plausible method.

2.8 Conclusion

We have proposed a method of information maximization used to simplify internal representations in terms of the number of hidden units. Information is defined by the decrease of uncertainty or entropy. If the information is increased, it can be predicted with more certainty which input pattern is given into input units. Inversely, we can easily estimate which hidden unit is strongly activated by input patterns.

If information is maximized, only one hidden unit is turned on, while all the other hidden units are turned off. Thus, by maximizing the information, the number of hidden units is significantly reduced. Thus, information maximization can be used to condense the information on input patterns into a small number of hidden units.

One of the shortcomings inherent to information maximization is that generalization performance may be degraded by the generation of strongly negative connections. For prohibiting the generation of the negative connections, we have proposed modified information maximization methods. In the modified methods, an information maximizer has been introduced to maximize information. Thus, only the information maximizer generates strongly negative connections to achieve a high information state. This prohibition of the excessive generation of negative connections enables networks to have better generalization performance.

We have applied the information maximization method to the XOR problem, the symmetry problem and the inference of consonant cluster well-formedness in an artificial language. In the XOR problem, the number of hidden units is reduced to a minimum point. In the symmetry problem, the number of hidden units is explicitly reduced to a minimum number of hidden units with explicit symmetric input–hidden connections. Experimental results on artificial languages have clearly shown that the number of hidden units is significantly reduced and that explicit rules can be extracted by examining internal representations. For improved generalization performance, information must be maximized only by updating bias–hidden connections and all the other input–hidden connections should be eliminated by weight elimination and weight decay methods. Experimental results on the well-formedness problem show that generalization performance is significantly increased and the information can be increased.

We have focused upon the network sized reduction as an application of the information maximization method. As discussed in the previous section, the information maximization method has shown a possibility of describing many aspects of neural learning such as competitive learning and selectivity of responses in living systems.

Chapter 3

Rule Extraction From Trained Artificial Neural Networks

Alan B Tickle[1], Robert Andrews[1,3], Mostefa Golea[2] and Joachim Diederich[1]
[1] Queensland University of Technology
[2] Australian National University
[3] robert@fit.qut.edu.au

3.1 Introduction

It is becoming increasingly apparent that without some form of explanation capability, the full potential of trained artificial neural networks (ANNs) may not be realized. This chapter gives an overview of some of the more significant techniques developed to redress this situation. The discussion focuses on mechanisms, procedures, and algorithms designed to insert knowledge into ANNs (knowledge initialization), extract rules from trained ANNs (rule extraction), and utilize ANNs to refine existing rule bases (rule refinement). The chapter concludes with an examination of some of the current issues in rule extraction/refinement.

The rapid and successful proliferation of applications incorporating artificial neural network (ANN) technology in fields as diverse as commerce, science, industry, and medicine, offers a clear testament to the capability of the ANN paradigm. Of the three salient characteristics of ANNs which underpin this success, the first is the comparatively direct and straightforward manner in which artificial neural networks acquire information/'knowledge' about a given problem domain through a training phase. This process is quite distinct from the more complicated knowledge engineering/acquisition processes of symbolic AI systems. The second characteristic is the compact (albeit completely numerical) form in which the acquired information/'knowledge' is stored within the trained ANN and the comparative ease and speed with which this 'knowledge' can be accessed and used. The third characteristic is the robustness of an ANN solution in the presence of 'noise' in the input data. In addition to these characteristics, one of the most important advantages of trained artificial neural networks is the high degree of accuracy reported when an ANN solution is used to generalize over a set of previously unseen examples from the problem domain [209].

However, the success of the ANN paradigm is at a cost—an inherent inability to explain, in a comprehensible form, the process by which a given decision or output generated by an ANN has been reached. For artificial neural networks to gain an even wider degree of user acceptance and to enhance their overall utility as learning and generalization tools, it is highly desirable if not essential that an 'explanation' capability becomes an integral part of the functionality of a trained ANN. Such a requirement is mandatory if, for example, the ANN is to be used in what are termed 'safety-critical' applications such as airlines and power stations. In these cases it is imperative that a system user be able to validate the output of the artificial neural network under all possible input conditions. Further the system user should be provided with the capability to determine the set of conditions under which an output unit within an ANN is active and when it is not, thereby providing some degree of transparency of the ANN solution.

Apart from the direct contribution to enhancing the overall utility of artificial neural networks, the addition of an 'explanation' capability is also seen as having the potential to contribute to the understanding of how symbolic and connectionist approaches to artificial intelligence (AI) can be profitably integrated. It also provides a vehicle for traversing the boundary between the connectionist and symbolic approaches.

This chapter deals with the more significant techniques developed to date to extract rules from trained artificial neural networks and hence provide the requisite explanation capability.

Techniques discussed in this chapter will be those derived for application to general feed-forward multi-layered artificial neural network architectures utilizing a 'supervised' learning regime such as back-propagation as well as to a selection of recurrent ANNs. Techniques for rule extraction from 'specialized' artificial neural network learning architectures such as KBANN developed by Towell and Shavlik [96] and others are included in the discussion whereas techniques for rule extraction from architectures based on unsupervised learning are excluded.

3.2   The importance of rule extraction

Since rule extraction from trained artificial neural networks comes at a cost in terms of resources and additional effort, an early imperative in any discussion is to delineate the reasons why rule extraction is an important, if not mandatory, extension of conventional ANN techniques. The merits in including rule-extraction techniques as an adjunct to conventional artificial neural network techniques include the following.

3.2.1 Provision of a 'user explanation' capability

Within the field of symbolic AI the term 'explanation' refers to an explicit structure which can be used internally for reasoning and learning, and externally for the explanation of results to a user. Users of symbolic AI systems benefit from an explicit declarative representation of knowledge about the problem domain, typically in the form of object hierarchies, semantic networks, frames etc. The explanation capability of symbolic AI also includes the intermediate steps of the reasoning process, e.g. a trace of rule firings, a proof structure etc, which can be used to answer 'How' questions. Further, Gallant [151] observes that the attendant benefits of an explanation capability are that it also provides a check on the internal logic of the system as well as enabling a novice user to gain insights into the problem at hand.

Experience has shown that an explanation capability is considered to be one of the most important functions provided by symbolic AI systems. In particular, the salutary lesson from the introduction and operation of knowledge based systems is that the ability to generate even limited explanations (in terms of being meaningful and coherent) is absolutely crucial for the user acceptance of such systems. In contrast to symbolic AI systems, artificial neural networks have no explicit declarative knowledge representation. Therefore they have considerable difficulty in generating the required explanation structures. It is becoming increasingly apparent that the absence of an 'explanation' capability in ANN systems limits the realization of the full potential of such systems and it is this precise deficiency that the rule-extraction process seeks to redress.

While provision of an explanation capability is a significant innovation in the ongoing development of artificial neural networks, of equal importance is the 'quality' of the explanations delivered [155]. It is here that the evolution of explanation capabilities in symbolic AI offers some valuable lessons into how this task of extracting rules from trained artificial neural networks might be directed. For example practitioners in the field of symbolic AI have experimented with various forms of user explanation vehicles including in particular, rule traces. However, for some time it has been clear that explanations based on rule traces are too rigid and inflexible. Indeed one of the major criticisms of utilizing rule traces is that they always reflect the current structure of the knowledge base. Further, rule traces may have references to internal procedures (e.g. calculations), might include repetitions (e.g. if an inference was made more than once), and the granularity of the explanation is often inappropriate [155]. Perhaps one clear lesson from using rule traces is that the transparency of an explanation is by no means guaranteed. For example experience has shown that an explanation based on a rule traces from a poorly organized rule-base with perhaps hundreds of premises per rule could not be regarded as being 'transparent'.

A further example of the limitations of explanation capabilities in symbolic AI systems which should, if possible, be obviated in the extraction of rules from trained artificial neural networks, comes from Moore and Swartout [192]. They note that the early use of 'canned' text or templates as part of user explanations has been shown to be too rigid, that systems always interpret questions in the same way, and that the response strategies are inadequate. Further, although efforts have been made to take advantage of natural-language dialogues with artifices such as mixed initiatives, user models, and explicitly planned explanation strategies, there is little doubt that current systems are still too inflexible, unresponsive, incoherent, insensitive, and too rigid.

In summary while the integration of an explanation capability (via rule extraction) within a trained artificial neural network is crucial for user acceptance, such capabilities must if possible obviate the problems already encountered in symbolic AI.

3.2.2   Extension of ANN systems to 'safety-critical' problem domains

While the provision of a 'user explanation' capability is one of the key benefits in extracting rules from trained ANNs, it is certainly not the only one. For example within a trained artificial neural network the capability should also exist for the user to determine whether or not the ANN has an optimal structure or size. A concomitant requirement is for ANN solutions to not only be transparent as discussed previously but also for the internal states of the system to be both accessible and able to be interpreted unambiguously. Satisfaction of such requirements would make a significant contribution to the task of identifying and if possible excluding those ANN-based solutions that have the potential to give erroneous results without any accompanying indication as to when and why a result is sub-optimal.

Such a capability is mandatory if neural-network based solutions are to be accepted into a broader range of applications areas and in particular, 'safety-critical' problem domains such as air traffic control, the operation of power plants, medical surgery etc. Rule extraction offers the potential for providing such a capability.

3.2.3   Software verification and debugging of ANN components in software systems

A requirement of increasing significance in software-based systems is that of verification of the software itself. While the task of software verification is important it is also acknowledged as being difficult, particularly for large systems. Hence if artificial neural networks are to be integrated within larger software systems which need to be verified, then clearly this requirement must be met by the ANN as well. At their current level of development, rule-extraction algorithms do not allow for the verification of trained artificial neural networks, i.e. they do not prove that a network behaves according to some specification. However, rule-extraction algorithms provide a mechanism for either partially or completely 'decompiling' a trained artificial neural network. This is seen as a promising vehicle for at least indirectly achieving the required goal by enabling a comparison to be made between the extracted rules and the software specification.

3.2.4 Improving the generalization of ANN solutions

Where a limited or unrepresentative data set from the problem domain has been used in the ANN training process, it is difficult to determine when generalization can fail even with evaluation methods such as cross-validation. By being able to express the knowledge embedded within the trained artificial neural network as a set of symbolic rules, the rule-extraction process may provide an experienced system user with the capability to anticipate or predict a set of circumstances under which generalization failure can occur. Alternatively the system user may be able to use the extracted rules to identify regions in input space which are not represented sufficiently in the existing ANN training set data and to supplement the data set accordingly.

3.2.5 Data exploration and the induction of scientific theories

Over time neural networks have proven to be extremely powerful tools for data exploration with the capability to discover previously unknown dependences and relationships in data sets. As Craven and Shavlik [138] observe, 'a (learning) system may discover salient features in the input data whose importance was not previously recognized'. However, even if a trained artificial neural network has learned interesting and possibly non-linear relationships, these relationships are encoded incomprehensibly as weight vectors within the trained ANN and hence cannot easily serve the generation of scientific theories. Rule-extraction algorithms significantly enhance the capabilities of ANNs to explore data to the benefit of the user.

3.2.6 Knowledge acquisition for symbolic AI systems

One of the principal reasons for introducing machine learning algorithms over the last decade was to overcome the so-called 'knowledge acquisition' problem for symbolic AI systems [202]. Further, as Sestito and Dillon [205] observe, the most difficult, time-consuming, and expensive task in building an expert system is constructing and debugging its knowledge base.

The notion of using trained artificial neural networks to assist in the knowledge acquisition task has existed for some time. An extension of these ideas is to use trained artificial neural networks as a vehicle for synthesizing the knowledge that is crucial for the success of knowledge-based systems. Alternatively domain knowledge which is acquired by a knowledge engineering process may be used to constrain the size of the space searched during the learning phase and hence contribute to improved learning performance.

The necessary impetus for exploring these ideas further could now come from two recent developments. The first is a set of recent benchmark results such as those of Thrun *et al* [208] where trained artificial neural networks have been shown to outperform symbolic machine learning methods. The second is from developments in techniques for extracting symbolic rules from trained artificial neural networks which could be directly added to the knowledge base.

3.3 Problem overview

Having identified the importance of rule extraction to the continued development and success of the ANN paradigm, the next step is to provide a succinct expression of the problem to be addressed. A useful starting point is a basic recognition that within a trained artificial neural network, knowledge acquired during the training phase is encoded as:

(i)   The network architecture itself (e.g. the number of hidden units).
(ii) An activation function associated with each (hidden and output) unit of the ANN.
(iii) A set of (real-valued) numerical parameters (called weights).

In essence the task of extracting explanations (or rules) from a trained artificial neural network is therefore one of interpreting in a comprehensible form the collective effect of the three points above.

An ancillary problem to that of rule extraction from trained ANNs is that of using the ANN for the 'refinement' of existing rules within symbolic knowledge bases. Whereas the rule-extraction process normally commences with an empty symbolic rule base, the starting point for the rule-refinement process is some initial knowledge about the problem domain expressible in the form of symbolic rules. A crucial point, however, is that the initial set of rules may not necessarily be complete or even correct [157]. Irrespective of the quality of the initial rule base, the goal in rule refinement is to use a combination of ANN learning and rule-extraction techniques to produce a 'better' (i.e. a 'refined') set of symbolic rules which can then be applied back in the original problem domain. In the rule refinement process, the initial rule base (i.e. what may be termed 'prior knowledge') is inserted into an ANN by programming some of the weights. (In this context, 'prior knowledge' refers to all of the production rules known prior to commencement of the ANN training phase.) The rule-refinement process then proceeds in the same way as normal rule extraction viz:

(i)   Train the network on the available data set(s).
(ii) Extract (in this case the 'refined') rules—with the proviso that the rule-refinement process may involve a number of iterations of the training phase rather than a single pass.

3.4   A classification scheme for rule-extraction algorithms

Andrews *et al* [120] suggested an overall taxonomy for categorizing the wide variety of techniques available for extracting rules from trained artificial neural networks. The taxonomy was based on a total of five primary criteria, viz:

(i)   The expressive power of the extracted rules.
(ii)   The translucency of the view taken by the technique of the underlying ANN units.
(iii) A measure of the portability of the rule-extraction technique across various ANN architectures, i.e. the degree to which the underlying ANN incorporates specialized training regimes.
(iv)   The quality of the extracted rules.
(v)   The algorithmic complexity of the rule-extraction/refinement technique.

The following is a brief discussion of each of the above criteria.

3.4.1   The expressive power of the extracted rules

The primary dimension of the proposed classification scheme labelled as 'the expressive power of the extracted rules' focuses directly on the actual output presented to the end user from the rule-extraction/rule-refinement process viz. the rules themselves. Classifying rule-extraction techniques according to this criteria has resulted in the formation of three classes of algorithm. Firstly there is a large group where rules are expressed using conventional (i.e. two-valued Boolean) symbolic logic in the form 'if...then...else...'. Secondly there is the group of algorithms that express the knowledge embodied in the ANN using concepts drawn from 'fuzzy' logic. A characteristic of this group is that they use set membership functions to deal with what are called 'partial truths'. For example a fuzzy rule is of the form if x is low and y is high then z is medium; where low, high, and medium are fuzzy sets with corresponding membership functions. A third group of algorithms expresses rules in forms other than the above. Examples here are the techniques for extracting finite-state automata that describe regular grammars from recurrent networks [156].

   At this point in time there are no techniques where the extracted rules are represented in first-order logic form, i.e. rules with quantifiers and variables.

3.4.2   The translucency of the underlying ANN units

This dimension of the classification schema is based on the view of the granularity of the underlying ANN assumed by the rule-extraction algorithm. At the maximum level of the granularity spectrum the ANN is viewed by the rule-extraction algorithm as a set of individual hidden and output units, and rules are extracted which describe the behaviour of individual hidden and output units. Such techniques were termed 'decompositional' by Craven and Shavlik

decompositional   *pedagogical*

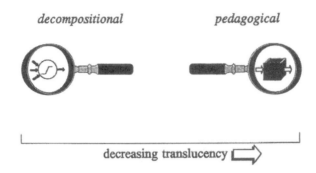

Figure 3.1. The translucency spectrum.

[138] and this terminology was adopted by Andrews *et al* in their survey paper [120]. At the other end of the granularity spectrum are those methods which view the underlying ANN as a 'black box' with the focus then on finding rules that map ANN inputs (attribute/value pairs) directly to outputs (membership of, or exclusion from some target class). Such techniques are termed 'pedagogical'. The degree to which the rule-extraction algorithm 'looks inside' the ANN is termed the translucency of the algorithm (see figure 3.1).

'Decompositional' approaches treat each unit in the ANN as an isolated entity and hence yield rules in which the antecedents (inputs to the unit) and consequents (unit output) are expressed in terms that are local to the unit from which they are derived. A process of aggregation is then applied to transform these local rules into a composite rule base for the ANN as a whole. Decompositional algorithms extract rules by employing some method of determining one or more combinations of incoming links whose summed weights guarantee the unit's bias is exceeded regardless of the activation value present on other incoming links. Methodologies employed to find these appropriate combinations of incoming links include exhaustive search and test, heuristically limited search and test, or direct decompilation of weights to rules.

The distinguishing characteristic of pedagogical techniques is that the rule extraction task is viewed 'as a learning task where the target concept is the function computed by the network and the input features are simply the network's input features' [138]. Where such techniques are used in conjunction with a symbolic learning algorithm, the basic motif is to use a generate and test method to create a set of cases/examples; utilize the generalization capability of the trained ANN to attach a classification to each example/case; and then use the symbolic learning algorithm to derive the rule set.

In addition to these two main categories of rule-extraction techniques Andrews *et al* also propose a third category which they labelled as 'eclectic' to accommodate those rule-extraction techniques which incorporate features of both the decompositional and pedagogical motifs. Membership in this category is assigned to techniques which utilize knowledge about the internal architecture and/or weight vectors in the trained ANN to complement a symbolic learning algorithm.

3.4.3 The portability of the rule-extraction algorithm

This classification dimension refers to the extent to which a given rule-extraction algorithm is generally applicable across different ANN architectures. Of the rule-extraction techniques considered by Andrews *et al* the majority were purpose built to suit a specific network architecture. Thus for these techniques to be applied to a given problem the relevant ANN had first to be trained on the problem. Portability is a highly desirable characteristic as it allows the rule-extraction technique to be applied *ipso facto* to existing, trained ANNs already *in situ*.

3.4.4 Quality of the extracted rules

The quality of extracted rule sets was first considered by Towell and Shavlik [96] in which they proposed the following criteria for evaluating rule quality:

(i)   Accuracy.
(ii) Fidelity.
(iii) Consistency.
(iv) Comprehensibility.

In this context a rule set is considered to be accurate if it can correctly classify previously unseen examples. Similarly a rule set is considered to display a high level of fidelity if it can mimic the behaviour of the artificial neural network from which it was extracted by capturing all of the information embodied in the ANN. An extracted rule set is deemed to be consistent if, under differing training sessions, the artificial neural network generates rule sets which produce the same classifications of unseen examples. Finally the comprehensibility of a rule set is determined by measuring the size of the rule set (in terms of the number of rules) and the number of antecedents per rule.

From the work of Giles and Omlin [157] additional rule-quality criteria which are pertinent to the ancillary problem of rule refinement are that the overall process must preserve genuine knowledge/rules, and correct wrong prior information/rules.

In passing it should be observed that in many 'real world' problem domains the simultaneous optimization of all multiple evaluation criteria may not always be desirable. For example in 'safety-critical' applications it is imperative that the artificial neural network be validated under all possible input conditions. This may create a situation for example where rule comprehensibility is sacrificed for rule accuracy and fidelity.

3.4.5   Complexity of the rule-extraction algorithm

The final dimension in the classification scheme described by Andrews *et al* is that of the algorithmic 'complexity' of a given rule extraction procedure. The inclusion of such a dimension reflects an almost universal requirement for the algorithms underpinning the rule-extraction process to be as efficient as possible. Decompositional techniques that employ a 'search and test' strategy to determine the conditions under which a given hidden or output unit will be active face the problem that the search space is exponential in the number of inputs to the node. Thus some heuristics must be introduced to limit the search space. Pedagogical techniques that use a 'generate and test' strategy face a similar problem in that suitable heuristics must be used to obviate the need for enumerating all possible examples in the problem space.

3.5   Rule-extraction techniques

3.5.1   Decompositional approaches to rule extraction

The earliest reported techniques adopted a decompositional approach to the extraction of rules from ANNs trained using standard back-propagation as the training regime. Conventional Boolean rules were extracted at the level of the individual (hidden and output) units within the trained ANN [127]. Of particular interest are the KT algorithm described by Fu [149] and the algorithm described by Towell and Shavlik [96] in which the basic motif is to search initially for sets of weights containing a single link/connection of sufficient (positive) value to guarantee that the bias on the unit being analysed is exceeded irrespective of the values on the other links/connections. If a link is found which satisfies the criterion, it is written as a rule. The search then proceeds to subsets of two elements *et seq.* and the rules extracted at the individual unit level are then aggregated to form the composite rule base for the ANN as a whole. A schematic of the basic algorithm as reported by Towell and Shavlik is given below.

- For each hidden and output unit:

  - Extract up to SR subsets of the positively weighted incoming links for which the summed weight is greater than the bias on the unit.

- For each element p of the SR subsets:

  - Search for a set SN of a set of negative attributes so that the summed weights of p plus the summed weights of $N - n$ (where N is the set of all negative attributes and n is an element of N) exceed the threshold on the unit.

- With each element n of the SN set, form a rule: 'if p and NOT n, then the concept designated by the unit'.

The first implementation of this style of algorithm was the KT algorithm developed by Fu [149]. In this implementation the problem of mapping the output from each (hidden and output) unit into a Boolean function was achieved by the simple artifice viz.

$$if\ 0 \leq output \leq threshold\_1 \Rightarrow no,\ if\ threshold\_2 \leq output \leq 1 \Rightarrow yes$$

$$where\ threshold\_1 < threshold\_2$$

Fu reported initial success in applying the KT algorithm to the problem domain of wind shear detection by infrared sensors.

A more recent example of this line of approach is the Subset algorithm developed by Towell and Shavlik. In their implementation, the artificial neural network is constructed in a way such that the computed value of the activation function in each hidden and output unit is either 'near' a value of one (i.e. 'maximally' active) or 'near' a value of zero (i.e. 'inactive'). Hence links carry a signal equal to their weight or no signal at all. They showed that their Subset implementation is capable of delivering a set of rules which are, at least 'potentially', tractable and smaller than many handcrafted expert systems. However, a major concern with both the KT and Subset algorithms is that the solution time for finding all possible subsets is a function of the size of the power set of the links to each unit, i.e. the algorithm is exponential. One option used by Fu for restricting the size of the solution search space is to place a ceiling on the number of antecedents per extracted rule. Unfortunately this potentially has adverse implications for rule quality since some rules may be omitted. Notwithstanding their limitations, the inherent simplicity of this class of algorithms still makes them extremely useful devices for explaining the mechanics of rule extraction. It also offers the capability to provide transparency of the trained ANN solution at the level of individual hidden and output units.

An important development in the utilization of specialized ANN architectures, was the publication of the *M*-of-*N* algorithm which is one component of the total KBANN package utilized by Towell and Shavlik. The *M*-of-*N* concept is a means of expressing rules in the form: 'If (*M* of the following *N* antecedents are true) then ...'. Towell and Shavlik cite as one of the main attractions of this approach a natural affinity between these rules and the 'inductive bias' of artificial neural networks. The phases of this algorithm are shown below:

(i)   Generate an artificial neural network using the KBANN system and train using back-propagation. With each hidden and output unit, form groups of similarly weighted links.
(ii) Set link weights of all group members to the average of the group.
(iii) Eliminate any groups which do not significantly affect whether the unit will be active or inactive.
(iv) Holding all link weights constant, optimize biases of all hidden and output units using the back-propagation algorithm.

(v) Form a single rule for each hidden and output unit; the rule consists of a threshold given by the bias and weighted antecedents specified by the remaining links.

(vi) Where possible, simplify rules to eliminate superfluous weights and thresholds.

This algorithm addresses the crucial question of reducing the complexity of rule searches by clustering the ANN weights into equivalence classes and then explicitly searching for rules of the form: 'If (M of the following N antecedents are true) then...'. For example, with reference to the unit shown in figure 3.2, the four extracted rules could be written as the single rule: 'If 3 of (B, C, D, not E) then A'.

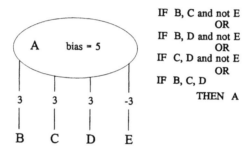

Figure 3.2. Rules extracted from a unit.

Towell and Shavlik use two dimensions in the knowledge based artificial neural network (KNN), for assessing the quality of rules extracted both from their own algorithm and from the set of algorithms they use for the purposes of comparison:

(i) The rules must accurately categorize examples that were not seen during training.

(ii) The extracted rules must capture the information contained.

In their view this method inherently yields a more compact rule representation than conventional conjunctive rules produced by algorithms such as Subset. In addition the algorithm outperformed a subset of published symbolic learning algorithms in terms of the accuracy and fidelity of the rule set extracted from a cross-section of problem domains.

3.5.2 Decompositional algorithms that directly decompile weights to rules

The direct decompilation of weights to rules obviates the need to involve exhaustive search and test strategies in the rule-extraction algorithm and thus makes the algorithm computationally efficient. In order that direct decompilation be possible a meaning relevant to the problem domain must be able to be ascribed to:

• Each hidden and output unit of the ANN.
• Each weight of each hidden and output unit.

Local function networks such as radial basis function (RBF) networks [191] with a single hidden layer of basis function units, perform function approximation and classification by mapping a local region of input space (hypercube or hyperellipsoid) directly to an output. The two conditions above can be met by either constraining the network such that at most one hidden unit exhibits appreciable activation in response to an input pattern, or by including in the extracted rule a 'belief value' or 'certainty factor' which indicates the degree to which the individual unit contributed to the output.

Under these circumstances individual hidden units can be decompiled to form a rule of the form: 'IF the input lies in the hypercube represented by the hidden unit THEN consequent, represented by the hidden unit output, is TRUE.' If the basis functions are allowed to 'overlap' (i.e. more than one basis function is allowed to be active in classifying an input vector), this necessitates the use of the certainty factor in the rules extracted from the network. By contrast the RULEX algorithm constrains the underlying rapid back-propagation (RBP) network such that the local functions do not overlap. The local functions used in the RBP network are constructed using pairs of axis parallel sigmoids. The local response region is created by subtracting the value of one sigmoid from the other. Geva and Sitte [153] described a parametrization and training scheme for networks composed of such sigmoid based hidden units and showed how these networks can be structured to facilitate rule extraction.

In the ith dimension, the sigmoids are parameterized according to centre, c_i, breadth, b_i, and edge steepness, k_i.

The pairs of sigmoids are given by the two equations:

$$U_i^+ = \frac{1}{1 + e^{(-(x_i - c_i + b_i)k_i)}}$$

$$U_i^- = \frac{1}{1 + e^{(-(x_i - c_i + b_i)k_i)}}$$

where x is the input vector, c is the reference vector, b represents the effective widths of each ridge, and k represents the edge steepness of each ridge.

The combination $U_i^+ - U_i^-$ forms a ridge parallel to the axis in the ith dimension (see figures 3.3 and 3.4). The intersection of N such ridges forms a local peak at the point of intersection but with secondary ridges extending away to infinity on each side of the peak (see figure 3.5). These ridges can be 'cut off' by the application of a suitable sigmoid to leave a locally responsive region (see figure 3.6). The activation V for this sigmoid is given by:

$$V = \frac{1}{1 + e^{-(\sum(U_i^+ - U_i^-) - B)K}}$$

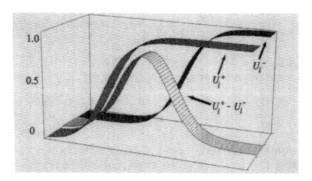

Figure 3.3. Two sigmoids.

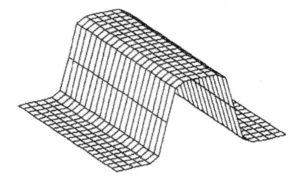

Figure 3.4. Sigmoids combined to form a ridge.

where B is set to the dimensionality of the input domain and K is set in the range 4–8. The network output O is:

$$O = \sum_{\mu=1}^{N} V_\mu w_\mu.$$

An incremental, constructive training algorithm is used with training, (for rule extraction), involving adjusting, by gradient descent, the centre, c_i, breadth, b_i, and edge steepness, k_i, parameters of the sigmoids that define the local response units. The output weight, w, is held constant. In classification problems where the desired output is either 0 or 1 this measure forces the bumps to avoid overlapping.

The parameters that define the local response units of the RBP network can be decompiled into rules of the form:

$$IF \ \forall 1 \leq i \leq n : x_i \in [x_i \ lower, x_i \ upper]$$

$$THEN \ Pattern \ Belongs \ to \ the \ Target \ Class.$$

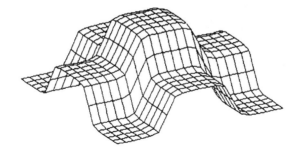

Figure 3.5. Intersection of ridges.

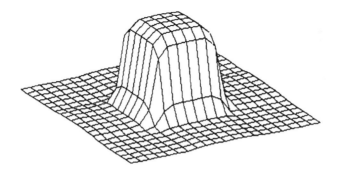

Figure 3.6. Locally responsive region.

Here x_i *lower* represents the lower limit of activation of the ith ridge and x_i *upper* represents the upper limit of activation of the ith ridge. These values can be calculated from the ridge parameters as:

$$x_i \ lower = c_i - b_i - \frac{\beta}{k_i}$$

$$x_i \ upper = c_i + b_i + \frac{\beta}{k_i}$$

where:

$$\beta = \ln\left(\frac{\ln(3)}{K - \ln(3)}\right).$$

RULEX is suitable for both continuous data and discrete data. RULEX also has facilities for reducing the size of the extracted rule set to a minimum number of propositional rules. This is achieved by removing redundant antecedent conditions, use of negations in antecedents, merging rules into more general rules, and by removing redundant rules.

Berthold and Huber [126] structure their rectangular basis function (RecBF) networks in such a way that there is a one-to-one correspondence between hidden units and rules. RecBF networks consist of an input layer, a hidden layer of RecBF units, and an output layer with each unit in the output layer representing a class. The hidden units of RecBF networks are constructed as hyper-rectangles with their training algorithm derived from that used to train RBF networks.

The hyper-rectangles are parametrized by a reference vector, r, which gives the centre of the rectangle, and two sets of radii, $\lambda_*^{+,-}$, which defines the core rectangle, and $\Lambda_*^{+,-}$, which describes the support rectangle (see figure 3.7).

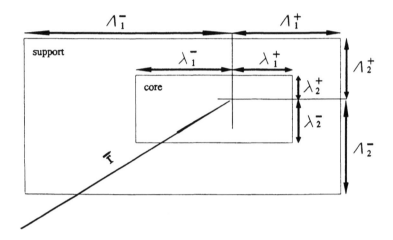

Figure 3.7. A RecBF unit.

The core rectangle includes data points that definitely belong to the class and the boundary of the support rectangle excludes data points that definitely do not belong to the class, i.e. the support rectangle is just an area where there are no data points. $R(*)$, the activation function for a RecBF unit is:

$$R(x) = \min_{1 \le i \le n} A(x_i, r_i, \lambda_i^-, \lambda_i^+)$$

where x represents the input vector, r represents the reference vector of the unit, and s is a vector representing individual radii in each dimension and $A(*)$ is the Signum activation function:

$$A(x_i, r_i, \sigma_i) = \{1 : r_i - \lambda_i^- \le x_i \le r_i + \lambda_i^+\} \{0 : else\}.$$

Training the RecBF network is by the dynamic decay algorithm, (DDA) [126]. This algorithm is based on three steps:

- *Covered.* A new training point lies inside the support rectangle of an existing RecBF. Extend the core rectangle of the RecBF to cover the new point

- *Commit.* A new pattern is not covered by a RecBF of the correct class. Add a new RecBF with centre the same as the training instance and widths as large as possible to avoid overlapping any existing RecBF.
- *Shrink.* A new pattern is incorrectly classified by an existing RecBF. The RecBF's widths are shrunk so that the conflict is resolved.

The main difference between RBF basis functions and RecBF basis functions is that the RecBF hyper-rectangles have finite radii in each input dimension. This allows straightforward interpretation of the RecBF parameters as rules of the form:

$$IF \ \forall \ 1 \leq i \leq n : x_i \in [r_i - \lambda_i^-, r_i - \lambda_i^+] \subset (r_i - \Lambda_i^-, r_i - \Lambda_i^+)$$
$$THEN \ Class \ c.$$

Here $[r_i - \lambda_i^-, r_i - \lambda_i^+]$ represents the core-rectangle region of the RecBF unit and $(r_i - \Lambda_i^-, r_i - \Lambda_i^+)$ represents the support-rectangle region of the RecBF unit.

Rules of this form have a condition clause for each of the n dimensions of the problem domain. This reduces the comprehensibility of the extracted rule set by including rules which contain antecedents for don't care dimensions, i.e. dimensions which the network does use to discriminate between input patterns. Don't care dimensions are those where $r_i - \lambda_i^- \leq x_{i\,min}$ and $r_i - \lambda_i^+ \geq x_{i\,max}$, where $x_{i\,min}$ is the smallest possible allowable value of the ith input dimension and $x_{i\,max}$ is the largest possible allowable value of the ith input dimension. Using the above scheme, condition clauses for don't care dimensions are removed from the rules extracted from RecBF networks.

One problem associated with RecBF networks is that the DDA training algorithm trains to zero error on the training set. This can result, in the case of noisy data sets or problem domains not suited to description by hyper-rectangles, in a network solution that has a low error rate but which is overtrained on the data set. This sort of solution will compromise the comprehensibility of the extracted rule set by producing rules that describe exceptions, i.e. the noisy data points. To combat this Berthold and Huber use a pruning strategy to reduce the rule set. Rules are pruned until an acceptable compromise is reached between classification accuracy and rule set size.

3.5.3 Pedagogical approaches to rule extraction

One of the earliest published 'pedagogical' approaches to rule extraction is that of Saito and Nakano [202]. In this implementation the underlying artificial neural network is treated as a 'black box' with rules from a medical diagnostic problem domain being extracted from changes in the levels of the input and output units. Saito and Nakano also deal with the problem of constraining the size of the solution space to be searched by avoiding meaningless combinations of inputs (i.e. medical symptoms in this problem domain) and restricting the maximum number of coincident symptoms to be considered. Even with these heuristics in place, the number of rules extracted on a relatively simple problem domain was exceedingly large. This result highlights one of the major concerns with rule-extraction techniques viz. that the end product is explanation and not obfuscation.

The VI-analysis (VIA) technique developed by Thrun [209] is also the epitome of a 'pedagogical' approach in that it extracts rules that map inputs directly into outputs. The algorithm uses a generate-and-test procedure to extract symbolic rules from standard back-propagation artificial neural networks which have not been specifically constructed to facilitate rule extraction. The basic steps in the procedure are shown below.

3.5.4   The VIA algorithm

(i)   Assign arbitrary intervals to all (or a subset of all) units in the ANN. These intervals constitute constraints on the values for the inputs and the activations of the output.

(ii)   Refine the intervals by iteratively detecting and excluding activation values that are provably inconsistent with the weights and biases of the network.

(iii)   The result of the previous step is a set of intervals which are either consistent or inconsistent with the weights and biases of the network. (In this context an interval is defined as being inconsistent if there is no activation pattern whatsoever which can satisfy the constraints imposed by the initial validity intervals.)

Thrun likens the approach to sensitivity analysis in that it characterizes the output of the trained artificial neural network by systematic variations in the input patterns and examining the changes in the network classification. The technique is fundamentally different from other techniques which analyse the activations of individual units within a trained ANN in that focus is on what are termed 'validity intervals'. A validity interval of a unit specifies a maximum range for its activation value. Establishing the validity intervals in which a unit in the trained artificial neural network becomes active involves solving a linear programming problem and hence the algorithmic complexity is dependent on the particular linear programming algorithm selected.

The author notes that one of the salient characteristics of the VIA algorithm is the capability to constrain the size of the rule search space by allowing the validity of more general rules to be determined before specific rules are examined. The resultant technique provides a generic tool for checking the consistency of rules within a trained ANN. The VIA algorithm is designed as a 'general purpose' rule-extraction procedure. Thrun uses a number of examples to illustrate the efficacy of his VIA technique including the XOR problem, the 'three monks' problem(s) and a robot arm kinematics (i.e. continuously valued domain) problem. While the VIA technique does not appear to be limited to any specific class of problem domains Thrun reports that VIA failed to generate a complete set of rules in a relatively complex problem domain involving the task of training a network to read aloud (NETtalk).

The 'rule-extraction-as-learning' approach of Craven and Shavlik [138] is another significant development in rule-extraction techniques utilizing the 'pedagogical' approach . The core idea is to 'view rule extraction as a learning task where the target concept is the function computed by the network and the input features are simply the network's input features'. A schematic outline of the overall algorithm is shown below.

- Initialize rules for each class.
- For each class c, $R_c := 0$.
- Repeat:
  - $e := Examples()$.
  - $c := Classify(e)$.
  - If e not covered by R_c then learn a new rule:
    * $r :=$ conjunctive rule formed from e
    * For each antecedent r_i of r:
      · $\acute{r} := r$ but with r_i dropped
      · If $Subset(c, \acute{r}) = True$ then $r := \acute{r}$
    * $R_c := R_c \vee r$
- Until stopping criterion met.

The role of the Examples function is to provide training examples for the rule-learning algorithm. The options used are:

(i) Select members of the set used for training the artificial neural network.
(ii) Random sampling.
(iii) Random creation of examples of a specified class, as outlined below.
     Random creation of examples algorithm in rule-extraction-as-learning:

- Create a random example, for each feature e_i with possible values $v_{i1}, \ldots, v_{in} : e_i := randomly\_select(v_{i1}, \ldots, v_{in})$.
- Calculate the total input s to output unit (which has a threshold value θ).
- If $s \geq \theta$ then return e.
- Impose random order on all feature values (considering the values in order), for each value v_{ij}:
  - If changing feature e_i's value to v_{ij} increase s:
    * $e_i := v_{ij}$
  - If $s \geq \theta$ then return e.

Craven and Shavlik use a function which they call *Subset* to determine whether the modified rule still agrees with the network, i.e. if all instances that are covered by the rule are members of the given class. A salient characteristic of this technique is that depending on the particular implementation used, the rule-extraction-as-learning approach can be classified either as a 'pedagogical' or 'decompositional'. The key is in the version of the *Subset* procedure used to establish whether a given rule agrees with the network. This procedure accepts a class label c and a rule r, and returns true if all instances covered by r are classified as members of class c by the network. If, for example, Thrun's VIA algorithm (as discussed previously) is used for this procedure then the approach is 'pedagogical' whereas if an implementation such as that of Fu [149] is used the classification of the technique is 'decompositional'. As with the VIA technique discussed earlier, the rule-extraction-as-learning technique does not require a special training regime for the network. The authors suggest two 'stopping criteria' for controlling the rule-extraction algorithm viz:

(i)   Estimating whether the extracted rule set is a sufficiently accurate model of the ANN from which the rules have been extracted.
(ii)   Terminating after a certain number of iterations have resulted in no new rules (i.e. a 'patience' criterion).

The authors report both on the algorithmic complexity of the technique as well as the quality of the extracted rules (with particular emphasis on rule 'fidelity' which is measured by comparing the classification performance of a rule set to the trained artificial neural network from which the rules were extracted. The fidelity of a rule set is the fraction of examples on which the rule set agrees with the trained artificial neural network.

The BRAINNE system of Sestito and Dillon [205] is also designed to extract rules from an artificial neural network trained using standard back-propagation. In this context it has been classified as pedagogical since the basic motif is to use a measure of the closeness between the network's inputs and outputs as the focal point for generating the rule set. The further classification of this approach as one requiring a specialized ANN training regime is based on their novel idea of taking an initial trained network with m inputs and n outputs and transforming it into a network with $m + n$ inputs (and n outputs). This transformed network is then retrained. The next phase in the process is to perform a pair-wise comparison of the weights for the links between each of the original m input units and the set of hidden units with the weights from each of the n additional input units and the corresponding hidden units. The smaller the difference between the two values, the greater the contribution of the original input unit (i.e. an attribute from the problem domain) to the output. A major innovation in the BRAINNE technique is the capability to deal with continuous data as input without first having to employ a discretizing phase. The BRAINNE technique both automatically segments the continuous data into discrete ranges and extracts corresponding if . . . then . . . else . . . rules directly. The authors report success in applying the BRAINNE technique to a 'real world' problem domain of interpreting submarine sonar data.

3.5.5 Eclectic rule-extraction techniques

In addition to the two main categories of rule-extraction techniques (viz. decompositional and pedagogical), Andrews *et al* also propose a third category which they labelled as eclectic. This category was designed to accommodate those rule-extraction techniques which incorporate elements of both the decompositional and pedagogical approaches. In reality the eclectic category is a somewhat diffuse group. This can be illustrated by the fact that one prominent example of an eclectic technique is the variant of rule-extraction-as-learning technique of Craven and Shavlik which utilizes the (decompositional) KT algorithm of Fu in the role of determining whether a given rule is consistent with the underlying ANN network.

Another example of the eclectic approach to ANN knowledge elicitation is the DEDEC approach of Tickle *et al* [210]. Essentially, DEDEC extends into the domain of knowledge elicitation from trained ANNs the work on identifying ANN causal factors [152], reducts [196], and functional dependences [154] as a precursor to rule extraction.

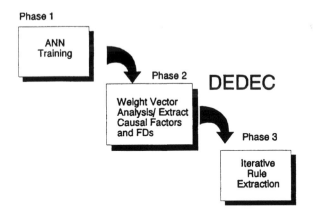

Figure 3.8. Schematic outline of the three basic phases of the DEDEC approach.

Figure 3.8 shows an overall schematic outline of the three basic phases of the DEDEC approach. DEDEC is designed to be applicable across a broad class of multilayer feedforward ANNs trained using the back-propagation technique and the starting point is phase 1 in which an ANN solution to a given problem domain is synthesized. The intermediate phase (phase 2) is where the task of identifying a set of causal factors and functional dependences is performed. Hence it is this phase which distinguishes DEDEC from other pedagogically based rule-extraction techniques and which provides the basis for assigning the DEDEC to the eclectic rule-extraction category.

To date two different weight vector analysis techniques have been developed and applied in DEDEC phase 2. The first of these utilizes an existing algorithm [152] for determining causal factors in a trained ANN based on calculating the relative weight shares of the ANN inputs. Within DEDEC phase 2 this basic algorithm has been adapted and extended for use in a broad range of ANN architectures including cascade correlation and an implementation of a 'local response' ANN network [153]. In addition to the weight sharing technique, DEDEC also incorporates a coefficient reduction approach to identifying causal factors and functional dependences in a trained ANN using an adaptation of an algorithm which was originally designed to be used in conjunction with linear programming problems.

For the DEDEC approach, the final phase (phase 3) is essentially the learning or pedagogical phase. It comprises a set of basic algorithms and techniques for eliciting the requisite sets of symbolic rules by learning from a selected set of cases generated by the trained ANN using the causal factor/functional dependence information extracted at phase 2.

3.6   Extraction of fuzzy rules

Parallel to the development of techniques for extracting Boolean rules from trained artificial neural networks has been the synthesis of corresponding techniques for extracting fuzzy rules (neurofuzzy systems). Analogous to the techniques discussed previously for conventional Boolean logic systems, typically, neurofuzzy systems comprise three distinct elements. The first is a set of mechanisms/procedures to insert existing expert knowledge in the form of fuzzy rules into an artificial neural network structure (i.e. a knowledge initialization phase). The essential difference here is that this step involves the generation of representations of the corresponding membership functions. The second element is the process of training the ANN which, in this case, focuses on tuning the membership functions according to the patterns in the training data. The third element in the process is the analysis and extraction of the refined knowledge embedded in the form of a set of modified membership functions. Horikawa *et al* [168] observe that the identification of the initial set of fuzzy inference rules to be modelled has proven to be a difficult task as have attempts at simultaneously undertaking the tasks of rule identification and membership tuning.

One of the earliest works in this area was that of Masuoka *et al* [183] who used a decompositional approach to refine an initial set of fuzzy rules extracted from experts in the problem domain. The technique incorporates a specialized three-phase ANN architecture. In the input phase a three-layer artificial neural network comprising an input unit, one or two hidden units, and an output unit was used to represent the membership function of each rule antecedent (i.e. the input variables). The fuzzy operations on the input variables (e.g. AND, OR, etc) are represented by a second distinct phase labelled as the rule net (RN) phase and the membership functions which constitute the rule consequents are represented in a third (output) phase using the same motif as for the input phase. In this technique the problem of eliciting a compact set of rules as the output is tackled by pruning at the RN phase those connections in the network which are less than a threshold value.

In a similar vein Berenji [125] demonstrated the use of a specialized artificial neural network to refine an approximately correct knowledge base of fuzzy rules used as part of a controller. (The problem domain selected in this case was a cart–pole balancing application.) The salient characteristic of this technique is that the set of rules governing the operation of the controller is known and the ANN is used to modify the membership functions both for the rule preconditions and the rule conclusions.

Horikawa *et al* [168] developed three types of fuzzy neural network which can automatically identify the underlying fuzzy rules and tune the corresponding membership functions by modifying the connection weights of the ANNs using the back-propagation algorithm. In this approach, the initial rule base is created either by using expert knowledge or by selectively iterating through possible combinations of the input variables and the number of membership functions. The fuzzy neural network model FuNe I developed by Halgamuge and Glesner [163] generalizes this work by using a (rule based) process to initially identify 'rule relevant nodes for conjunctive and disjunctive rules for each output'. Halgamuge reports on the successful application of the FuNe I technique to a benchmark problem involving the classification of Iris species as well as three real world problems involving the classification of solder joint images, underwater sonar image recognition, and handwritten digit recognition.

The fuzzy-MLP model of Mitra [184] specifically addresses the problem of providing the end user with an explanation (justification) as to how a particular conclusion has been reached. Here the set of rule antecedents is determined by analysing and ranking the weight vectors in the trained ANN to determine their relative influence (impact) on a given output (class). The fuzzy-MLP model has been applied to the medical problem domain of diagnosing hepatobiliary disorders.

Okada *et al* [193] incorporated elements of knowledge initialization, rule refinement (via the tuning of membership functions), and rule extraction in a fuzzy inference system incorporating a seven-layer structured ANN. In this implementation, two layers of the model are used to provide representations of the membership functions for the input variables (presented in a separate input layer) and another layer is used to represent membership functions for the rule consequents. Separate layers are also used to construct the rule antecedents (incorporating mechanisms for supporting fuzzy logical operations) and rule consequents. The authors report a significant improvement in prediction accuracy of the model in comparison with a conventional three-layer neural network in the application problem domain of financial bond rating.

Fuzzy ARTMAP developed by Carpenter and Tan [130] is another example of a situation in which a highly effective rule-extraction algorithm has been designed to work in conjunction with a specific supervised learning ANN, i.e. the Fuzzy ARTMAP system. The algorithm is decompositional because a characteristic feature of the Fuzzy ARTMAP system is that each (category) node roughly corresponds to a rule. Furthermore, the weight vector associated with each node can be directly translated into a verbal or algorithmic description of the rule antecedents. This is in contrast to a 'conventional' back-propagation network where the role of hidden and output units in the total classification process is not usually as explicit.

3.7   Techniques for performing rule refinement

The first successful method for prestructuring an artificial neural network such that the classification behaviour of the network was consistent with a given set of propositional rules was Towell and Shavlik's KBANN algorithm. KBANN is in essence a domain theory refinement system. A initial approximately correct domain theory is provided as a propositional rule base, the KBANN network is created from the rules and then trained with examples drawn from the problem domain. Finally a refined set of *M*-of-*N* rules are extracted from the trained network. Figure 3.9 below shows the process of converting a rule-based knowledge base (a), to the corresponding hierarchical representation of the knowledge base (b) and to a knowledge based neural network (KNN) that represents the rule base (c).

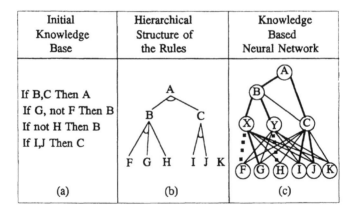

Figure 3.9. Construction of a KNN by the KBANN algorithm.

In (c) the solid and dotted lines represent necessary and prohibitory dependences respectively. The KNN shown in (c) results from the translation of the knowledge base. Units X and Y in (c) are introduced to handle the disjunction in the rule set. Otherwise each unit in the KNN corresponds to a consequent or an antecedent in the knowledge base. The thick lines in (c) represent heavily weighted links in the KNN that correspond to dependences in the knowledge base. The thin lines in (c) represent links added to the network to allow refinement of the knowledge base.

The core requirements of the KBANN/M-of-N approach are:

(i) The requirement for either a 'rule set' to initialize the ANN or a special training algorithm that uses a 'soft-weight sharing' algorithm to cluster weights.
(ii) The requirement for a special network training regime.
(iii) The requirement for hidden units to be approximated as threshold units (this is achieved by setting the parameter s in the activation function $1/[1+e^{-sx}]$ to be greater than a value of 5.0).
(iv) The requirement that the extracted rules use an intermediate term to represent each hidden unit.

This gives rise to the concern that the approach may not enable a sufficiently accurate description of the network to be extracted. It is also worth noting that one of the basic tenets of the M-of-N approach is that the meaning of a hidden unit in the artificial neural network generated as part of the initialization process, does not change during the training process. Given that M-of-N is essentially a rule-refinement system this may be true in general and, in fact, Towell and Shavlik report empirical confirmation from trained ANNs in their study. However, in the case where the meaning of a unit does change during training, the comprehensibility of the extracted rules may be significantly degraded.

The M-of-N approach has been tested successfully on a diverse range of problem domains including two from the field of molecular biology viz. the promoter recognition problem and the splice-junction determination problem. Towell and Shavlik also undertook a detailed comparison of the quality of the rules extracted using their technique with other ANN rule-extraction techniques as well as symbolic learning techniques.

Local response networks are inherently suitable for rule refinement. Firstly, it is conceptually easy to see how a local response unit can be converted to a symbolic rule. In all cases this conversion is achieved by describing the area of response of the individual units in terms of a reference vector that represents the centre of the unit and a set of radii that determine the effective range of the unit in each input dimension, (and hence the boundaries of the unit). The rule associated with the unit is formed by the conjunct of these effective ranges in each dimension. Rules extracted from each local response unit are thus propositional and of the form:

$$IF \ (x_1 \in [x_{1\,min}, x_{1\,max}]) \wedge (x_2 \in [x_{2\,min}, x_{2\,max])} \ \ldots$$

$$\wedge (x_n \in [x_{n\,min}, x_{n\,max}])$$

$$THEN \ pattern \ belongs \ to \ class \ c$$

where $[x_{1\,min}, x_{1\,max}]$ represents the effective range in the ith input dimension.

Secondly, because each local response unit can be described by the conjunct of some range of values in each input dimension it makes it easy to add units to the network during training such that the added unit has a meaning that is directly related to the problem domain. In networks that employ incremental learning schemes (such as RBP networks) a new unit is added when there is no significant improvement in the global error. The unit is chosen such that its reference vector, i.e. the centre of the unit, is one of the as yet unclassified points in the training set. Thus the premise of the rule that describes the new unit is the conjunction of the attribute values of the data point with the rule consequent being the class to which the point belongs.

This also makes local function networks suitable for rule refinement. As long as the knowledge to be used for network initialization can be stated in the form shown above, the starting configuration of the network is given by forming a local response unit for each rule in the knowledge base.

A further advantage of local response units over symbolic methods in particular is that when continuous values are quantized (e.g. $[1, 10] \rightarrow \{1, 2, \ldots, 10\}$) the local response units will tend to generalize over a sub-range if there are no conflicting data in the middle. For example, a ridge on a value of say 3, can extend to cover the range 3–7 even if the data do not contain evidence for some value in between. Symbolic methods such as decision trees leave 'holes' and need pruning to recover generalization. This property is apparent in the RBP networks where the local function ridges will grow through an area of no conflicting data only establishing boundaries when patterns that don't belong to the target class are encountered.

A technique can be based on the premise that prior knowledge of the problem domain is available in the form of a set of rules. An ANN $y = NN(x)$, which makes a prediction about the state of y given the state of its input x can be instantiated as a set of basis functions, $b_i(x)$, where each basis function describes the premise of the rule that results in prediction y. The degree of certainty of the rule premise is given by the value of $b_i(x)$ which varies continuously between 0 and 1. The rule conclusion is given by $w_i(x)$ and the network architecture is given as:

$$y = NN(x) = \frac{\sum_i w_i(x)b_i(x)}{\sum_j b_j(x)}.$$

If the w_i are constants and the basis functions chosen are multivariate Gaussians (i.e. individual variances in each dimension) the above equation reduces to the network described by Moody and Darken [215], who show how the basis functions can be parametrized by encoding simple logical if–then expressions as multivariate Gaussians. For instance the rule:

$$IF\ [(x_1 \approx a)\ AND\ (x_4 \approx b)]\ OR\ (x_2 \approx c)\ THEN\ y = dx^2$$

is encoded as:

$$premise : b_i = \exp\left(-\frac{1}{2}\frac{(x_1 - a)^2 + (x_4 - b)^2}{\sigma^2}\right) + \exp\left(-\frac{1}{2}\frac{(x_2 - c)^2}{\sigma^2}\right)$$

$$conclusion : w_i(x) = dx^2$$

Training can proceed in any of four modes including:

(i)   Forget, where training data is used to adapt NNinit by gradient descent (i.e. the sooner training stops, the more initial knowledge is preserved.
(ii) Freeze, where the initial configuration is frozen (i.e. if a discrepancy between prediction and data occurs, a new basis function is added).
(iii) Correct, where a parameter is penalized if it deviates from its initial value.
(iv) Internal teacher, where the penalty is formulated in terms of the mapping rather than in terms of the parameters.

Classification is performed by applying Bayesian probability and making the assumption that $P(x|\text{class}_k)P(\text{class}_k) \approx \sum_i b_{ik}(x)$ to obtain:

$$P(\text{class}_k|x) = \frac{P(x|\text{class}_k)P(\text{class}_k)}{\sum_l P(x|\text{class}_l)P(\text{class}_l)}.$$

Rule extraction is performed by directly decompiling the Gaussian (centre: μ_{ij}, width: δ_{ij}) pairs to form the rule premise and attaching a certainty factor, w_{ij} to the rule. After training is complete, a 'pruning' strategy (rule refinement), is employed to arrive at a solution which has the minimum number of basis functions (rules), and the minimum number of conjuncts for each rule. The strategy is shown below.

- While Error < Threshold

  - Either prune/remove basis function which has least importance to the network (remove the least significant rule).
  - Or prune conjuncts by finding the Gaussian with the largest radius and setting this radius to infinity (effectively removing the associated input dimension from the basis function).
  - Retrain the network till no further improvement in error.

- End While

Our latest research, RULEIN, is a procedure for turning a propositional if–then rule into the parameters that define a RBP local response unit. This process involves determining from the rule the active range of each ridge in the unit to be configured. This means setting appropriately the upper and lower bounds of the active range of each ridge, $x_{i\,lower}$ and $x_{i\,upper}$, and then calculating the centre, breadth, and steepness parameters (c_i, b_i, k_i) according to the equations given below.

Setting $x_{i\,lower}$ and $x_{i\,upper}$ appropriately involves choosing values such that they 'cut off' the range of antecedent clause values. For discriminating ridges, i.e. those ridges that represent input pattern attributes that are used by the unit in classifying input patterns, these required values will be those that are mentioned in the antecedent of the rule to be encoded. For non-discriminating ridges the active range can be set to include all possible input values in the corresponding input dimension. (Non-discriminating ridges will be those that correspond to input pattern attributes that do not appear as antecedent clauses of the rule to be encoded.)

The ridge centre c_i, can be calculated as:

$$c_i = \frac{x_{i\,upper} - x_{i\,lower}}{2}.$$

Now b_i is calculated from the centre, c_i, $x_{i\,lower}$, the initial ridge steepness, K_0, and the unit output weight K, and is given as:

$$b_i = \frac{K_0(c_i - x_{i\,lower})}{K_0 - \beta}$$

where the value for β is given above.

The steepness parameter k_i can be calculated as:

$$k_i = \frac{K_0}{b_i}.$$

After network training has taken place the refined rules can be extracted using RULEX.

3.8   Rule refinement and recurrent networks

Work in the area of rule refinement and recurrent networks has centred on the ability of recurrent networks to learn the rules underlying a regular language, (where a regular language is the smallest class of formal languages in the Chomsky hierarchy). Giles and Omlin [158] state that a regular language is defined by a grammar $G = \langle S, N, T, P \rangle$ where:

- $S =$ start symbol.
- $N =$ non-terminal symbol.
- $T =$ terminal symbol.
- $P =$ a production of the form $AA \rightarrow a$ or $A \rightarrow aB$ where $AB \in N$, and $a \in T$.
- The regular language generated by G is denoted $L(G)$.

Giles and Omlin also discuss the equivalence between the regular language L and the deterministic finite-state automaton (DFA) M which acts as an acceptor for the language $L(G)$, i.e. DFA M accepts only strings which are members of $L(G)$. They formally define a DFA as a 5-tuple $M = \langle S, Q, R, F, \delta \rangle$ where:

- $S = \{a_1, \ldots, a_m\}$ is the alphabet of the language L.
- $Q = \{q_1, \ldots, q_n\}$ is a set of states.
- $R \in Q$ is a start state.
- $F \subseteq Q$ is a set of accepting states.
- $\delta : Q \times \sum \rightarrow Q$ defines state transitions in M.

Acceptance of a string x by the DFA M is defined as the DFA M reaching an accepting state after being read by M. Acceptance of the string x by the DFA M implies that x is a member of the regular language $L(M)$, (and hence also of the regular language $L(G)$).

Cleeremans *et al* [136] and Elman [28] showed that recurrent networks were capable of being trained such that the behaviour of the trained network emulated a given DFA. Cleeremans *et al* concluded that the hidden unit activations represented past histories and that clusters of these activations represented the states of the generating automaton. Giles and Omlin [157] extended the work of Cleeremans and described a technique for extracting complete deterministic finite-state automata from second-order, dynamically driven recurrent networks which is described below.

The networks used by Giles and Omlin have N recurrent hidden units labelled S_j, K special non-recurrent input units labelled I_k, and $N^2 \times K$ real-valued weights labelled W_{ijk}. The values of the hidden neurons are referred to collectively as state vectors S in the finite N-dimensional space $[0, 1]^N$. (Second order is taken to mean that the weights W_{ijk} modify a product of the hidden (S_j) and input (I_k) neurons which allows a direct mapping of {state, input}$\Rightarrow$ {next} and means the network has the representational potential of at least finite state automata.) Their method for extracting DFA from recurrent networks is given below:

(i)   Divide the output of each of the N state neurons into q intervals, (quantization levels). This results in q^N partitions of the hidden state unit space.

(ii) Starting in a defined initial network state generate a search tree with the initial state as its root and the number of successors of each node equal to the number of symbols in the input alphabet. (Links between nodes correspond to transitions between DFA states.)

(iii) Perform (breadth first) a search of the tree by presenting all strings up to a certain length in alphabetical order starting with length 1. Make a path from one partition to another. When:

  –   (a) A previously visited partition is reached, then only the new transition is defined between the previous and the current partition, i.e. no new DFA state is created and the search tree is pruned at that node.

  –   (b) An input causes a transition immediately to the same partition, then a loop is created and the search tree is pruned at that node.

(iv) Terminate the search when no new DFA states are created from the string set initially chosen and all possible transitions from all DFA states have been extracted.

(v) For each resulting path, if the output of the response neuron is greater than 0.5 the DFA state is accepting; otherwise the DFA state is rejecting.

The extracted DFA depends on:

- The quantization level, q, chosen. Different DFAs will be extracted for different values of q.
- The order in which strings are presented (which leads to different successors of a node visited by the search tree).

Giles and Omlin state that these distinctions are usually not significant as they employ a minimization strategy which guarantees a unique, minimal representation for any extracted DFA. Thus DFAs extracted under different initial conditions may collapse into equivalence classes.

Rule insertion for known DFA transitions is achieved by programming some of the initial weights of a second-order recurrent network with N state neurons. The rule-insertion algorithm assumes $N > N_s$ where N represents the number of neurons in the network and N_s is the number of states in the DFA to be represented.

In the recurrent networks used by Giles and Omlin the network changes state S at time $t + 1$ according to the equations:

$$S_i^{(t+1)} = g(\Xi_i)$$

$$\Xi_i = \sum_{j.k} W_{ijk} S_j^{(t)} I_k^{(t)}$$

where g is a sigmoid discriminant function.

To encode a known transition $\delta(s_j, a_k) = s_i$ Giles and Omlin arbitrarily identify DFA states s_j and s_i with state neurons S_j and S_i respectively. This transition can be represented by having S_j have a high output (≈ 1) and S_i have a low output (≈ 0) after the input symbol a_k has entered the network via input neuron I_k.

Setting W_{ijk} to a large positive value will ensure that $S_i^{(t+1)}$ will be high, and setting W_{ijk} to a large negative value will ensure that $S_j^{(t+1)}$ will be low. All other weights are set at small random values.

To program the response neuron to indicate whether the resulting DFA state is an accepting or rejecting state the weight W_{0ie} is set large and positive if s_i is an accepting state, and large and negative if s_i is a rejecting state (where e is a special symbol that marks the end of an input string.)

Network training proceeds after all known transitions are inserted into the network by encoding the weights according to the above method. After training, the refined rules/DFA is extracted using the method described above. Giles and

Omlin conclude that network initialization reduces training time and improves generalization on the example problems studied.

The method proposed by Giles and Omlin is an alternative to the method described by Frasconi *et al* [148] for injecting finite state automata into first-order recurrent radial basis function (R^2BF) networks. R^2BF networks consist of two layers; a locally tuned processing unit layer and a sigmoidal unit layer termed the state layer. Feedback connections exist from the state layer to the radial basis function layer (see figure 3.10).

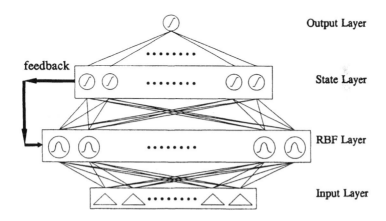

Figure 3.10. The R^2BF network architecture.

The authors also describe a penalty function that is applied to the state layer neurons which forces the output of the state neurons to approximate Boolean values and biases the network towards finite state behaviour resulting in the learned automata states being represented by very small 'clusters' in the network state space. Prior knowledge about the problem domain is injected into the network by programming the radial basis function centres and widths to represent the minterms of the canonical form of the next-state function. After training the learned automata can be extracted according to the process described below:

(i)   Compute all the state vectors for each sequence of the learning set.
(ii) $K \leftarrow 2$.
(iii) Use the k-mean algorithm for partitioning the set of state vectors into K subsets.
(iv) If the distance between two centres is less than d_c then $K \leftarrow K + 1$. Go to third step above.
(v)   Compute the transition table:

   –   (a) Since each cluster corresponds to a state of the automaton, for each cluster use its centre as network state t (or as initial condition), and from this state feed the network with one symbol.

- (b) Get the resulting state vector and find the cluster which corresponds to this vector.
- (c) Repeat these two steps for all symbols and states.

(vi) Compute the initial state by finding the cluster containing the initial network state.

(vii) Compute the accepting states:

- (a) Initialize the state neurons with each cluster (i.e. state of the automaton) and get the value of the output neuron.
- (b) If it is greater than 0.5 then the current cluster corresponds to an accepting state.

Frasconi demonstrates the technique successfully on the task of inductive inference of regular grammars.

3.9   Current issues in rule extraction and refinement

The field of rule extraction from trained artificial neural networks has achieved a degree of maturity. Researchers have moved beyond merely describing new techniques and reporting empirical results to formulating benchmark standards and developing a consistent theoretical base. The following is a discussion of some of the issues raised by recent theoretical work.

3.9.1   Limitations imposed by inherent algorithmic complexity

The survey paper of Andrews *et al* used algorithmic complexity of the rule-extraction algorithm as one of their criteria for categorizing rule-extraction processes. This criterion was introduced due to the observations of several authors, notably Fu, Towell and Shavlik, Thrun and more recently Viktor *et al* [99], that for certain 'real world' problem domains there exist potential problems due to the algorithmic complexity of various implementations of the rule-extraction process. Typically these problems relate to the algorithm either requiring a long time to find the maximally general solution or producing a large number of rules (with a consequent loss of comprehensibility). Further such authors have usually shown how a variety of heuristics may be employed to achieve a balance between solution time/effort, fidelity (degree to which the rule set mimics the underlying ANN) and accuracy (measure of the ability of the rule set to classify previously unseen examples from the problem domain) of the rule set, and comprehensibility of the rule set.

Golea [160] identified issues relating to the intrinsic complexity of the rule-extraction problem. The two key results were that, in the worst case, extracting the following are all NP-hard problems:

- The minimum disjunctive normal form (DNF) expression from a trained (feedforward) ANN.

- The best monomial rule from a single perceptron within a trained ANN.
- The best M-of-N rule from a single perceptron within a trained ANN.

3.9.2 Limitations on achieving simultaneously high accuracy, high fidelity, and high comprehensibility

Recently, rule-extraction techniques have been applied in an increasingly diverse range of problem domains. This increased exposure has also brought to light a potential conflict in attempting to maximize simultaneously the fidelity, accuracy, and comprehensibility criteria for evaluating the quality of the rules extracted from a trained ANN. The general nature of this problem was described by Golea and may be illustrated as in figure 3.11. (Recall that rule accuracy is a measure of the capability of the extracted rules to classify correctly a set of previously unseen examples from the problem domain. Rule fidelity is a measure of the extent to which the extracted rules mimic the behaviour of the ANN from which they were drawn whereas rule comprehensibility is assessed in terms of the size of the extracted rule set and the number of antecedents per rule.)

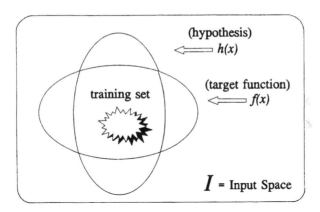

Figure 3.11. Training set, target function (f), and ANN solution (h).

Let I be the complete input space for a given problem domain and let D be a distribution defined on I where $D(x)$ represents the probability of seeing a given x in I. Let $f(x)$ be the function which actually maps or classifies a set of training cases drawn from the problem domain (i.e. the function which is the 'target' for the ANN training). In addition let $h(x)$ be the functional representation of the ANN solution and let R be the set of rules extracted from the ANN.

On the basis of the definitions presented above, the set of extracted rules R exhibits a high level of fidelity with respect to the ANN solution h if R can act as a surrogate for h (i.e. R and h can be used interchangeably). Hence a set of extracted rules R with a high level of fidelity will be as accurate as the ANN itself in classifying previously unseen examples from the problem domain. (In passing it should also be noted that a number of authors, e.g. Towell and Shavlik, have reported situations in which the extracted rule set exhibits a better generalization performance than the trained ANN from which the rule set was extracted.)

A standard measure of the accuracy of the ANN solution (h) (which results from the ANN training process involving the given training set) is given by the generalization error (e) in using h (i.e. the ANN solution) as a surrogate for the target function f viz.

$$e = \text{probability } \{x \in D \mid h(x) \neq f(x)\}.$$

Hence, if the generalization error e is small, a set of extracted rules R with a high level of fidelity will simultaneously exhibit a high level of rule accuracy (as defined above). Moreover, R can therefore act as a surrogate for the original target function f.

However, if the distribution D is localized in some region of the input space I then it is possible to synthesize an ANN solution h for which the generalization error e is small but for which h is neither equal to nor perhaps even close to f. Importantly, if the function h is significantly more complex than f, then the extracted rule set R will exhibit high levels of accuracy and fidelity but a correspondingly lower level of comprehensibility than the set of rules extracted, for example, by applying a symbolic induction algorithm directly to the data in the original training set.

The survey paper of Andrews *et al* listed one of the important benefits in extracting rules from a trained ANN as being the ability to identify situations in which certain regions of an input space were not represented sufficiently by data in an ANN training set. The view expressed in the survey was that this would enable the data set to be supplemented accordingly. The preceding discussion has highlighted the importance of this observation.

3.9.3   Rule extraction and the quality of ANN solutions

In most ANN architectures the initial values for the weight vectors which ultimately characterize an ANN solution are randomly assigned within the ANN training algorithm. Consequently the result of each separate instance of the ANN training is normally a unique ANN solution. After training the ANN solution is assessed in terms of:

- The size of the residual error on the training set.
- The size of the generalization error on the test or validation set.
- The number of hidden units required in the trained ANN.

Some trade-off between these three measures is often required particularly in situations involving noise in the training set. However, in most cases the implicit or in some cases explicit goal of the training phase is to arrive at an ANN solution with the minimum number of hidden units consistent with satisfying certain threshold criteria for the residual/generalization errors.

An issue which was not expanded upon in the survey paper of Andrews *et al* is the extent of the dependence of the efficacy of rule extraction techniques on the quality of the ANN solution (and by extension, the algorithm used in the ANN training phase). For example, because of their algorithmic complexity, the tractability of most decompositional approaches to rule extraction (and specifically those which involve some form of search process to find possible rule sets) is heavily dependent on the ANN having as close as practical to the minimum number of hidden units in the final configuration. More importantly, such decompositional techniques are critically dependent on each (hidden and output) unit possessing a separate and distinct meaning or representing a single concept or feature within the context of the problem domain.

However, a recent result by Bartlett [122] raises certain problems in this regard. In particular Bartlett has shown that the generalization error (e) of a trained ANN can be expressed in the form:

$$e \propto \frac{A^2}{n}$$

where n is the number of cases in the training set; and $\sum |w| \leq A$, i.e. the sum of the absolute values of the weights w for each (hidden and output) unit in the trained ANN is bounded by some positive constant A.

This result shows that an ANN solution may be found which exhibits good generalization behaviour (i.e. have acceptable low generalization error e) whilst being sub-optimal in terms of having the minimum number of hidden units. Moreover an important corollary is that an ANN solution could be found with good generalization capability in problems involving binary classifications but in which some or all of the hidden units do not possess a separate and distinct meaning or represent a single concept or feature within the context of the problem domain. As such the result has important implications for all rule-extraction techniques and in particular for those decompositional rule-extraction techniques for which this is a prerequisite.

3.9.4   Functional dependences, causal factors and rule extraction

To date almost the entire focus of the discussion regarding knowledge elicitation from trained ANNs has centred on the task of extracting as efficiently as possible a set of (symbolic) rules which explain the behaviour of the trained ANN. However, as was pointed out at the beginning of this chapter another reason for performing rule extraction is to discover previously unknown dependences and relationships in data sets. In particular a useful interim result from an end-user's point of view might simply be one that identifies which attributes, or combination of attributes, from the problem domain are the most significant (or alternatively the least significant) determinants of the decision/classification.

Moreover Holte [167] showed that for a broad range of data sets involving in some situations a large number of attributes, it is frequently possible to classify cases from the problem domain to an acceptable level of accuracy based on two or fewer dominant attributes.

A numerous and diverse range of techniques has been developed which are designed to enable the dominant attributes in a problem domain to be isolated. This range includes statistical techniques such as multiple regression analysis, discriminant analysis, and principal component analysis. Within the context of knowledge elicitation from trained ANNs, Garson [152] focused on the task of determining the relative importance of input factors used by the ANN to arrive at its conclusions. Garson termed these as causal factors between ANN inputs and outputs. In a similar vein Tickle *et al* [210] identified certain parallels between the process for identifying keys, superkeys, and in particular functional dependences [172] within the realm of relational database design. Specifically, for a given problem domain and a set of cases drawn from the problem domain (where each case comprises a set of attribute/value pairs), then in essence a functional dependence (FD) $X \rightarrow Y$ (read X determines Y) exists if the value of the attribute Y can be uniquely determined from the values of the attributes belonging to set X. More precisely the functional dependence $X \rightarrow Y$ is satisfied if for each pair of cases t_1 and t_2 in a given problem domain then $t_1[X] = t_2[X]$ (i.e. the set of attributes X for cases t_1 and t_2 are equivalent) it follows that $t_1[Y] = t_2[Y]$.

By way of illustration of the concept of functional dependences consider the set of cases $T = \{t_1 \ldots t_5\}$ involving attributes A, B, C and D shown in table 3.1 (from [172]). In this example the functional dependence A→C is satisfied because each case which has the same value of attribute A has the same value for attribute C. However, the functional dependence C→A is not satisfied because the cases t_4 and t_5 both have the value c_2 for attribute C (i.e. $t_4[C] = t_5[C]$ but $t_4[A] \neq t_5[A]$). In a similar way it can be shown that the functional dependence AB→D is satisfied.

Both the notion of a causal factor as introduced by Garson and the concept of a functional dependence as has been previously discussed can be viewed

Table 3.1. Functional dependences.

Case	A	B	C	D
t_1	a_1	b_1	c_1	d_1
t_2	a_1	b_2	c_1	d_2
t_3	a_2	b_2	c_2	d_2
t_4	a_2	b_3	c_2	d_3
t_5	a_3	b_3	c_2	d_4

as one in which a conceptual relationship is established between the domain attributes and the decision attribute(s). However, a functional dependence relies on having discrete values for the attributes whereas a causal factor as described by Garson can embrace both discrete and continuous valued attributes. Moreover the notion of a rule can be viewed as an extension of the functional dependence concept to the point of expressing a relationship between specific values of the domain attributes and the decision attribute(s) [154].

In any given problem domain there could exist numerous functional dependences. Hence in a relational database context attention is primarily focused on determining only what are termed left-reduced functional dependences, i.e. those which possess the property that while the functional dependence $X \rightarrow Y$ is satisfied, any proper subset $X \subset Y$ is not sufficient to determine Y. Applying these concepts in the realm of ANNs it is expected that a trained ANN would not necessarily reflect all of the possible functional dependences in a given dataset. This is because the intrinsic nature of the training process is to give prominence to those attribute/values or combination of attribute/values which leads to global error minimization.

In both the relational database context and also in the context of applying the functional dependence concept to knowledge elicitation from trained ANNs, the identification of the set of left-reduced functional dependences is important because the goal is to identify and eliminate superfluous/insignificant attributes. In addition, eliminating such attributes augurs well for ultimately determining a set of rules with the minimum number of antecedents [154].

As indicated previously in certain applications, the identification of causal factors and/or functional dependences may of itself provide considerable insight into the problem domain for the end-user. However, this may not be the only benefit. In the context of rule extraction from trained ANNs, one of the issues upon which comment has already been made is the complexity of the various rule-extraction algorithms. In particular algorithms such as KT and Subset are exponential in the number of ANN inputs. Hence reducing the number of attributes by eliminating those that are irrelevant in determining the decision has the potential of making a direct impact on the tractability of such algorithms by significantly reducing the search space [154]. In addition this has the potential to obviate one of the key problems in both the VIA and the rule-extraction-as-learning algorithms viz. finding an initial set of specific cases which can then be used to synthesize more general rules.

One impediment to the use of functional dependences to preprocess the training set data is that the determination of all functional dependences is in itself exponential in the number of domain attributes. Geva and Orlowski suggest some heuristics by which the process of discovering and enumerating all functional dependences may be made more tractable.

3.9.5   Extension to connectionist knowledge representation techniques

Andrews *et al* use the expressive power of the extracted rules as one of the criteria to classify rule-extraction algorithms. It was noted earlier in the chapter that the majority of the rule-extraction algorithms described to date produce either propositional or fuzzy rules. A more expansive view of the overall knowledge elicitation/rule-extraction task is taken by Dillon *et al* [142]. In particular the primary overall aim in their work is to form a link between the process of eliciting the knowledge embedded in a trained ANN and the more general issues of the knowledge representation and knowledge refinement [141]. As part of their approach Dillon utilizes a connectionist knowledge representation technique viz. the so called SHRUTI network which incorporates the notions of the dynamic binding of variables and reflexive reasoning. The role of the ANN in this approach is to use data from the problem domain to refine existing knowledge which has been used to initialize an ANN. Similarly the rules extracted initially from a trained ANN can be transformed into the representation of more general concepts within a SHRUTI network.

3.10   Conclusion

As evidenced by the diversity of 'real world' application problem domains in which rule-extraction techniques have been applied, there appears to be a strong and continuing demand for the end-product of the rule-extraction process viz. a comprehensible explanation as to how and why the trained ANN arrived at a given result or conclusion. This demand appears to fall broadly within two groups:

(i)   ANN solutions which have already been implemented and where *ipso facto* the user is interested in identifying and possibly exploiting the potentially rich source of information which already exists within the trained ANN.

(ii)   A 'green-field' situation where a user has a data set from a problem domain and is interested in what relationships exist both within the data given and what general conclusions can be drawn.

The first group requires the development of rule-extraction techniques which can be applied to existing ANNs. At this stage it would appear that, notwithstanding the initial success of 'decompositional' approaches such as that of the KT algorithm of Fu the 'pedagogical' approach is well placed to serve this set. Similarly it could be argued that the second group might well become the province of those rule-extraction techniques which use specialized artificial neural network training regimes, given the reported success of, for example, KBANN/M-of-N, BRAINNE, RULEX, etc. However, it is also clear that no single rule-extraction/rule-refinement technique or method is currently in a dominant position to the exclusion of all others.

A pressing problem then is the formulation of a set of criteria for matching the set of techniques to the requirements of a given problem domain. For example at a practical level, what has not yet emerged is a means of determining which rule-extraction technique is optimal for application problem domains involving real-valued data as distinct from discrete data. Further it is also uncertain as to whether the reported improvement in performance of ANN/rule-extraction techniques vis-à-vis other induction techniques for extracting rules from data, applies in all problem domains. Hence a pressing requirement is for a set of comparative benchmark results across a range of problem domains similar to that undertaken with the original 'three monks' problem proposed by Thrun.

A related issue is that in an increasing number of applications there are reports of situations in which the extracted rule-set has shown better generalization performance than the trained artificial neural network from which the rule-set was extracted. Similar observations have also been made in the area of extracting symbolic grammatical rules from recurrent artificial neural networks. However, Giles and Omlin also report that larger networks tend to show a poorer generalization performance. While these results are significant, what is not clear at this stage is the extent to which this superior performance can be ascribed to the elimination of the remaining error over the output unit(s) after the artificial neural network training has been completed (i.e. the 'rest' error). Hence an important research topic is also to identify the set of conditions under which an extracted rule set shows better generalization than the original network.

This chapter has described the reasons for the emergence of the fields of rule extraction and rule refinement from artificial neural networks and described a taxonomy for classification of rule-extraction algorithms. A selection of published rule-extraction/refinement techniques was discussed to illustrate the taxonomy. The chapter also highlighted a variety of important issues relevant to the field that deserve the attention of researchers in the field.

Acknowledgments

Some of this material is from R. Andrews, J. Diederich and A. B. Tickle. Survey and critique of techniques for extracting rules from trained artificial neural networks. *Knowledge Based Systems*, 8:373–389, 1995, reprinted with kind permission from Elsevier Science—NL, Sara Burgerhartstraat 25, 1055 KV Amsterdam, The Netherlands.

PART 2

NOVEL ARCHITECTURES AND ALGORITHMS

Many people working in the field of neural computing understand simple feed-forward networks, modelled in software using the backpropagation algorithm. However, new hardware and software architectures have been developed, both to increase the speed of neural networks and their accuracy in modelling a range of tasks. These new architectures should widen the appeal of neural computing. In this part of the book two novel hardware architectures are discussed, together with a description of fast training algorithms and the use of several networks in parallel.

Chapter 4

Pulse-Stream Techniques and Circuits for Implementing Neural Networks

Robin Woodburn[1] and Professor Alan F Murray
University of Edinburgh, UK
[1] robin.woodburn@ee.ed.ac.uk

4.1 Introduction

In this short tutorial guide to pulsed design, we have selected circuits for illustrative purposes that have underpinned some of our working neural chips. These are not necessarily the optimal circuits from our work. They have been chosen as case studies for their ease of explanation and to draw out important and useful concepts in pulsed design. We provide a set of references to some of the most up-to-date work in the pulsed area at the end of this chapter. Similarly, the treatment of pulsed circuit analysis that is presented here is pragmatic and simplified, rather than rigorous and complete. No important details are omitted or glossed over, but every reasonable corner has been cut!

4.2 Pulse-stream encoding

Pulse-stream signals encode analogue information in the time domain by modulating the *width* of a single pulse or the *frequency* of a stream of pulses, as shown in figure 4.1.

For a pulse-width modulated (PWM) signal, the width of the pulse represents an analogue value with a maximum value determined by the largest pulse-width and a minimum value determined by the narrowest pulse-width that is detectable. A PWM signal can carry out a computation within the time of the widest pulse (although there are likely to be computational overheads that add to this time). The maximum frequency of the signal therefore depends on the widest pulse.

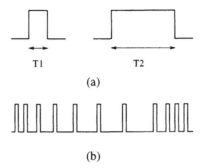

Figure 4.1. Encoding analogue information in pulse-stream signals: (a) a pulse-width modulated signal; (b) a pulse-frequency modulated signal.

A pulse-frequency modulated (PFM) signal uses fixed-width pulses that vary in frequency. The largest analogue value is represented by the maximum frequency, and the smallest by the minimum frequency. A PFM signal can take longer to carry out a computation than its PWM counterpart, because the PFM pulses have to be aggregated to establish the frequency of operation. There is, however, no restriction (in principle) on the minimum PFM signal.

Neither in the case of the PWM nor of the PFM signal is the amplitude of the signal modulated, because standard CMOS or TTL pulses are used. For neural applications, PFM signals have the appeal that they are closest in form to the asynchronous spiking of real neurons; however, most neural algorithms bear only a superficial resemblance to biological neural networks, so the decision to use PWM or PFM normally depends purely upon practical considerations.

Phase-encoding of information is also possible. We have not as yet used this method, although it is common in biological systems.

4.3   Basic neural computation on a chip

Commonly, artificial neural networks (ANNs) involve a limited class of computations. These comprise *multiply-and-add* computations and some *non-linearity*, usually a sigmoid function. For example, if the network is a multi-layer perceptron (MLP), at each node in the network a series of input states is multiplied by a set of weights and the sigmoid maps the summed results to a single output state. More formally, the output O_{pk} of the kth node for the pth input pattern is a sigmoid function σ of the sum of inputs O_{pkj} from the previous layer, weighted by the weights w_{kj}:

$$O_{pk} = \sigma_k \left(\sum w_{kj} O_{pkj} \right).$$

If we can create a circuit that will multiply the weights and states together, sum these products, and map them to the output state *via* a sigmoid function,

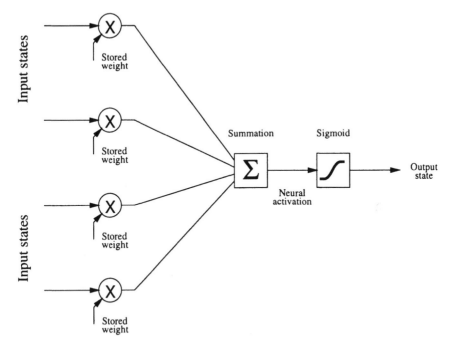

Figure 4.2. The typical sum-of-products operations of a neural network.

then we have all the constituents of a feed-forward ANN.

A block diagram of the feed-forward system would look like figure 4.2.

Our experience tells us that it is valuable to heed some basic principles in carrying out this task:

- There must be some electrical representation for input and output states. It makes sense to use pulsed signals at the chip boundary because pulses are robust. The obvious choice of signal to represent input and output states, then, is pulses.

- In analogue circuits, it is easy to add currents together, and this gives us a means of summing the products of a *weight × state* calculation. According to Kirchoff's current law: *the sum of the currents flowing into a node equals the sum of currents leaving the node*, as shown in figure 4.3. We should therefore like the output of each of the circuits that carries out the *weight × state* calculation to be a current.

- There must be an electrical representation for weights. There is something to be said for either voltage or current but our preferred choice is voltage.

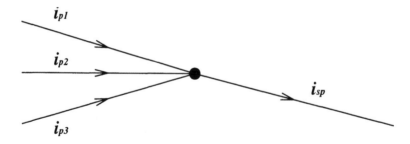

Figure 4.3. A single current i_{sp} can represent a sum of products when each of the product terms is represented in turn by a current i_p, so that $i_{p1} + i_{p2} + i_{p3} = i_{sp}$.

4.4   Computation using an analogue, two-quadrant multiplier

In this section, we design a two-quadrant multiplier to compute the *weight × state* product.   The emphasis here is on using simple ideas and hand calculations to give us a 'feel' for what is happening in the circuit and to calculate approximate component values. We then confirm our basic calculations using the circuit simulator, Hspice, to confirm the correctness of our analysis.

At its simplest, the pulsing circuit for a PWM signal can be reduced to a switch, and yet analogue computation can be performed. Figure 4.4 shows a two-quadrant, pulse-stream multiplier.

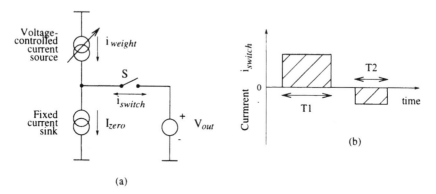

Figure 4.4.   (a) A schematic diagram of a two-quadrant, pulse-stream multiplier; (b) output pulses.

A voltage-controlled current source, which supplies a current i_{weight}, is connected in series with a current sink, which sinks a fixed, positive-value current, I_{zero}. A switch connects the current source and sink to a voltage source, V_{out}. While the switch S is closed, the current through the switch is:

$$i_{switch} = i_{weight} - I_{zero}.$$

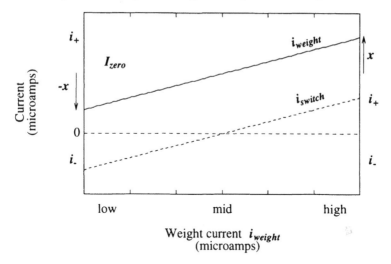

Figure 4.5. The relationship between the three currents i_{weight}, I_{zero} and i_{switch}.

The relationship between the currents in this circuit is depicted in figure 4.5. i_{weight} is allowed to vary over a range that is always positive in value but with its higher and lower values lying approximately symmetrically above and below I_{zero}.

As a result, i_{switch} varies in the same way as i_{weight}, but its mid-point is zero, that is, the current can take values in two quadrants.

Hence, the amplitude and polarity of the current through the switch is controlled by i_{weight}, and its duration is dependent on the time T the switch S is closed. In short, the circuit produces an output current of a magnitude, and sign, determined by i_{weight} and I_{zero}, in bursts or pulses whose width and frequency are determined by the switch S.

The area enclosed by the current pulse represents the amount of charge transferred. It also represents the product of the values of the *state* (encoded in time) and the *weight* (represented by the current amplitude):

$$i = \frac{dq}{dt}$$

$$\Rightarrow q = \int_0^T i \, dt$$

and, if the current has a constant value:

$$Q = IT.$$

We can calculate the total charge by integrating the current on a capacitor, and reading the value as a voltage:

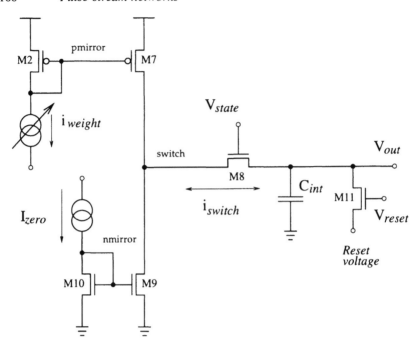

Figure 4.6. A circuit-diagram of the pulsed multiplier.

$$i = C\frac{dv}{dt}$$

$$\Rightarrow V = \frac{1}{C}\int_0^T i\,dt.$$

4.5 Designing a pulsed multiplier—a case study

We can realize this idea using transistors, as shown in figure 4.6 [17].

All the information we need to analyse the circuit is in table 4.1. The equations are only first-order approximations, and the values are also only approximate, but they allow us to do hand calculations on the circuit, which we can confirm later using Hspice simulations. (Remember when using these equations that, for an NMOS transistor, the drain/source terminals refer to the terminals connected to the higher/lower voltages while, for a PMOS transistor, the reverse is true.)

Each of the pairs of transistors M2/M7 and M10/M9 forms a current mirror. Transistors M2 and M10 are diode connected, and so inevitably fulfil the conditions, summarized in table 4.1, for operating in their saturation regions. The voltages on the gates of M2 and M10 are mirrored to M7 and M9,

Table 4.1. Equations for NMOS and PMOS transistors operating in saturation. The values for V_T and K are derived from one of the standard VLSI processes.

	NMOS	PMOS
Condition for saturation	$0 < V_{gs} - V_T \leq V_{ds}$	$0 < V_{sg} + V_T \leq V_{sd}$
Equations for operation in saturation region	$i_{ds} = \dfrac{\beta}{2}(v_{gs} - V_T)^2$	$i_{sd} = \dfrac{\beta}{2}(v_{sg} + V_T)^2$
	$v_{gs} = \sqrt{\dfrac{2i_{ds}}{\beta}} + V_T$	$v_{sg} = \sqrt{\dfrac{2i_{sd}}{\beta}} - V_T$
	where $\beta = \dfrac{KW}{L}$	
V_T (V)	0.82	−1.05
K (μA/V²)	65	21

respectively, and therefore so are the currents through the transistors. The currents are mirrored precisely only when the two transistors in a mirror-pair have the same drain–source voltage, for instance when $v_{ds_{M2}} = v_{ds_{M7}}$, but we can reduce inaccuracies in other circumstances by reducing the aspect-ratios of the transistors, so that they are long and thin.

Hence, M7 will act as a current source, i_{weight}, and M9 as a current sink, I_{zero}. Transistor M8 acts as the switch, being pulsed with a 0–5 V voltage V_{state}. Any charge being conducted through M8 is dumped onto, or drawn from, C_{int}.

So that we can make some initial calculations, let us give the variables in our circuit some actual values.

$$i_{zero} = 1 \; \mu A$$
$$i_{weight} = 0.4 \; \mu A\text{–}1.6 \; \mu A$$
$$\Rightarrow i_{switch_{max}} = 0.6 \; \mu A$$
$$i_{switch_{min}} = -0.6 \; \mu A.$$

For the transistors M2, M7, M9 and M10, we can set width and length:

$$W = 3 \; \mu m$$
$$L = 10 \; \mu m.$$

The voltage V_{out} changes as the charge on C rises and falls. When the switch is closed, that is when M8 is *on*, the node labelled *switch* in figure 4.6 changes to match. If the *switch* node-voltage rises or falls too far, then M7 and M9 will leave saturation mode and cease behaving as current sources. We can establish the range approximately using the information in table 4.1. Since the calculation is only a hand one, it will only be approximate, but we will be able to refine the approximation when we verify our calculations in simulation using Hspice.

To determine an acceptable range for the output voltage, we need to know when the conditions $V_{gs} - V_T = v_{ds}$ (for the NMOS current mirror) and $V_{sg} + V_T = v_{sd}$ (for the PMOS current mirror) are fulfilled. First, we need to know the voltage at the gates of transistors M9 and M7 in figure 4.6, using the equations from table 4.1.

For the NMOS transistor M9, the calculation is this:

$$V_{gs} = \sqrt{\frac{2i_{ds}}{\beta}} + V_T$$

$$= \sqrt{\frac{2i_{ds}L}{KW}} + V_T$$

$$= \sqrt{\frac{2 \times 1\ \mu\text{A} \times 10\ \mu\text{m}}{65\ \mu\text{A V}^{-2} \times 3\ \mu\text{m}}} + 0.82\ \text{V}$$

$$\approx 1.1\ \text{V}.$$

Knowing the gate voltage, we can then use the conditional equation from table 4.1 to determine the lowest voltage possible on v_d before the current mirror ceases to operate properly:

$$V_{gs} - V_T = v_{ds}$$
$$\Rightarrow v_d = V_g - V_T$$
$$= 1.08\ \text{V} - 0.82\ \text{V}$$
$$\approx 0.3\ \text{V}.$$

For the PMOS transistor M7, the calculation is similar. Using $i_{weight_{max}} = 1.6\ \mu\text{A}$:

$$v_{sg} = \sqrt{\frac{2i_{sd}}{\beta}} - V_T$$

$$= \sqrt{\frac{2i_{sd}L}{KW}} - V_T$$

$$= \sqrt{\frac{2 \times 1.6\ \mu\text{A} \times 10\ \mu\text{m}}{21\ \mu\text{A V}^{-2} \times 3\ \mu\text{m}}} + 1.05\ \text{V}$$

$$\approx 1.8\ \text{V}.$$

Hence, the maximum voltage on v_d is calculated so:

$$v_g = V_{DD} - v_{sg}$$

$$= 5.0 \text{ V} - 1.8 \text{ V}$$
$$= 3.2 \text{ V}$$
$$V_{sg} + V_T = v_{sd}$$
$$\Rightarrow v_d = V_g - V_T$$
$$= 3.25 \text{ V} + 1.05 \text{ V}$$
$$\approx 4.3 \text{ V}.$$

These calculations show that the range of voltages that the capacitor C can accommodate without the current-mirror transistors falling out of saturation is approximately 0.3–4.3 V, a range of 4 V. We note in passing that these conditions assume that M8 is an ideal switch, which of course it is not, although we can improve its characteristics by making it a transmission gate rather than a single, NMOS, transistor.

If we provide a reset voltage of 2.3 V, to represent a zero point, then the maximum voltage swing on v_{out}, in either a positive or negative direction, is 2 V. We can now calculate a suitable value for the capacitor C. If we set a maximum pulse width of 10 μs for V_{state} to represent a state input, and since $i_{switch_{max}} = 0.6 \ \mu$A, then:

$$C = \frac{Q}{V}$$
$$= \frac{IT}{V}$$
$$= \frac{0.6 \ \mu\text{A} \times 10 \ \mu\text{s}}{2 \text{ V}}$$
$$= 3 \text{ pF}.$$

4.6 Implementing weights

A topic not yet tackled is how to control the current i_{weight}. There is no 'correct' way to do this. One method proposed [17] is shown in figure 4.7.

There is insufficient space to explain the design of this circuit, but the principles on which it is based are easy to explain, and in any case this sort of circuit is well described in many standard textbooks. The transistors M1 to M5 form a low-gain, differential stage, driven by a tail-current mirror M6. A voltage v_{weight} on the gate of transistor M4 represents a weight, and the range of this weight-voltage, and the voltage V_{zero}, are chosen such that the current through M2 is a linear function of v_{weight}. The current through M2 is copied, in its turn, *via* the current mirror M2 forms with M7, as i_{weight}. Note that i_{weight} rises as v_{weight} rises, so that a low weight-voltage represents a low weight. The differential stage is designed such that a weight range of 2.4 V–4.0 V is converted to a current i_{weight} in the range 0.4 μA–1.6 μA, giving the correct

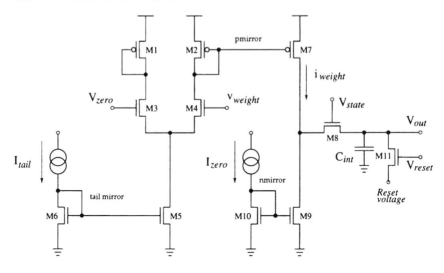

Figure 4.7. The synapse with a differential stage to convert an input weight-voltage v_{weight} into a current i_{weight}.

current input for the synapse. The current sources are provided off-chip by a power supply and large-value resistors.

4.7   Results from the multiplier circuit

We can test our analysis of the circuit in simulation using Hspice, and graphs of such tests are shown in figure 4.8.

Graph 4.8(a) is aimed at checking the maximum range possible at v_{out} while still enabling the current mirrors to operate effectively. The x-axis shows a sweep of v_{out} from the lower rail-voltage, 0 V to the higher rail-voltage, 5 V. The y-axis shows the current, i_{switch}, through the switch. The measurements are made for six values of current i_{weight}, from 0.4 μA to 1.6 μA. As we predicted with our earlier calculations, the correct value for the current through the switch degrades badly below 0.3 V and above 4.3 V but, inside this range, the values are close to correct. The 'horizontality' of the curves inside the range attests to the circuit's robustness to variations in v_{out}.

Graph 4.8(b) is a rather different way of looking at the same information, representing as it does the simulated version of the ideal curves shown in figure 4.5. Although the graph looks as though it comprises only three curves, these measurements have also been made for different values of v_{out}, and each graph is plotted 'on top' of the others. Again, the fact that there is such a small variation in the curves demonstrates the circuit's effectiveness.

The third graph, 4.8(c), is a demonstration of the whole circuit as it would be used in a feed-forward calculation. The x-axis represents different values of

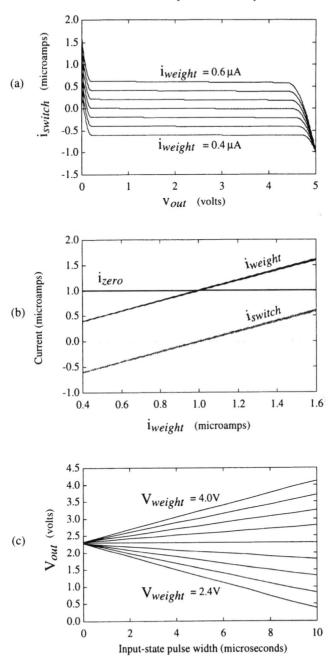

Figure 4.8. Graphs of circuit simulations. (a) The variation in i_{switch} as v_{out} varies. (b) The relationship of i_{weight}, i_{zero} and i_{switch}. (c) Final voltages on C_{int} for different values of weight-voltage and input-state pulses.

input state, that is pulses of varying widths, from 0 μs (no pulse at all) to 10 μs. The y-axis shows the final voltage on the integration capacitor C_{int} once the pulse has been applied.

To plot these curves, an input weight v_{weight}, say 2.4 V, is set on the weight input to the differential stage; the voltage on the capacitor C_{int} is reset to an initial voltage of 2.3 V; a pulse, say 0 μs, is applied; and the voltage on C_{int} is measured at the end of the pulse. The voltage on C_{int} is then reset and another pulse, say 1 μs, is applied and the final voltage on C_{int} measured. This process is repeated until the whole range of input pulses from 0 μs to 10 μs has been applied. A single curve can then be plotted on the graph. Now the input-weight-voltage is stepped up to, say, 2.6 V, another set of pulses is applied and measurements made, and a second curve is added to the graph. We continue the process until we have measurements for the full range of weights, from 2.4 V to 4.0 V.

The curves in figure 4.8(c) are evenly spaced and linear, and they converge to the same zero-point at *pulse-width* = 0 μs and v_{out} = 2.3 V, which is correct.

Bear in mind that these curves are from simulations only. In this discussion we have necessarily obscured many other issues that are important in achieving such results from a real chip, notably good matching of transistors, accurate sizing of capacitors, and interfacing with other circuitry.

4.8   Converting analogue outputs and inputs into pulses

Since analogue values are encoded as binary pulses, many of the advantages of analogue and digital signals can be realized in pulse-stream designs.

For example, we can 'read' the voltage on the activation capacitor using a simple comparator, and an inverted, double-sided, analogue ramp provided by a DAC, as shown in figure 4.9. As the ramp falls and rises again, the output of the comparator produces a pulse. The shape of the ramp can also be chosen to implement non-linear functions. For example, multi-layer perceptrons (MLPs) require the sigmoid function shown in figure 4.9. This function can be determined off-chip by generating an appropriate double-sided, non-linear ramp, so that the comparator output produces a pulse that is a function of the 'sigmoidal ramp'. The same principle can be applied to other functions such as the Gaussian function required by radial-basis-function (RBF) networks.

With the multiplier, op-amp, integrator and comparator circuits on a single chip, the inputs to the chip are analogue voltages (which can be stored digitally, off-chip), and pulses. With comparators on the chip at the inputs, we can even interface directly to analogue voltages. The outputs from the chip are pulses, which can be transmitted directly to other similar analogue chips or to digital circuits.

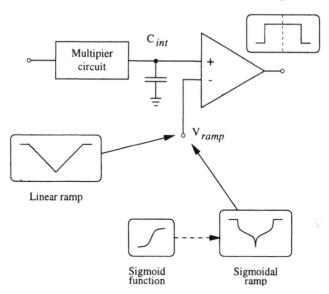

Figure 4.9. Using a comparator to convert voltages into pulses.

4.9 Centred pulses

We have chosen to modulate pulse widths symmetrically about a centre-line, by using double-sided ramps. This centring of pulses has several benefits. It reduces noise introduced by circuit switching and power surges on the chip, because the rising and falling edges of pulses of different widths occur at different times. Also, some computational functions are rendered very simple.

Figure 4.10 illustrates, as an example, how the value of one pulse might be subtracted from another using an XOR gate. The output is a series of short pulses that, provided the time-frame in which they occur is controlled, can be used in another computation. The sign of the subtraction can also be determined easily, using an SR flip-flop (which has been redesigned from the standard form so that it cannot settle in an indeterminate state). If the longer of the two pulses is applied to input S, Q is set; if the longer pulse is applied to R, Q is reset.

Naturally, if the asynchronous nature of a PFM approach is important to a particular application, pulse-centring is not an option.

4.10 Quantization

To what extent are time-encoded signals quantized in the systems we have described? With PWM, this depends on our means of providing pulses.

It may be, depending on the application, that it is critical to represent the state signals in an analogue fashion, and centred precisely, to allow accurate

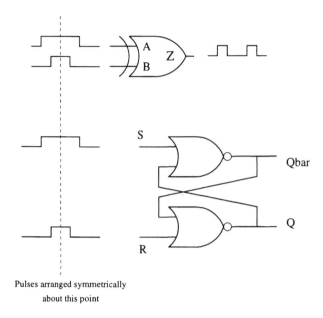

Pulses arranged symmetrically
about this point

Figure 4.10. 'Subtracting' pulses using an XOR gate.

calculation of, for example, the difference of two pulses or the sign of a pulse. If state pulses are provided using the comparator and an analogue, double-sided ramp, either for inputs to the chip or outputs from it, then the pulsed signals are truly analogue, and precisely centred.

PFM signals are also truly analogue, with the disadvantage that, to aggregate the pulses to carry out the computation will probably take considerably longer than the 'per-pulse' computation of PWM.

Where such accuracy is not important, and some level of quantization is tolerable, then for simplicity we might provide the ramp to the comparator using a DAC, which of course produces a stepped ramp, not an analogue ramp, and introduces quantization effects. Alternatively, we can usefully 'store' pulses in RAM, as shown in figure 4.11.

To 'fire' the pulses, we use a rapidly clocked counter to address successive locations, starting with address 0 and ending with the highest address in the RAM. Again, this introduces quantization effects, and there is the disadvantage that pulses made up of an odd number of addresses cannot be exactly centred on a mid-point. The level of quantization depends on the clock-speed; to reduce the quantization, the clock-speed can be increased. There are practical limits to the clocking-speed because, of course, to increase the clocking-speed means that, for a pulse of a given width, more RAM is required.

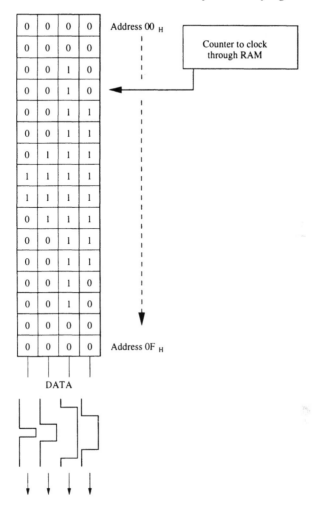

Figure 4.11. 'Firing' pulses by clocking through a RAM.

4.11 Process dependence of signals as VLSI technologies are scaled down

All the designs presented here were fabricated on VLSI processes that used 5.0 V and 0.0 V supply-rails. For example, the current-pulses described in section 4.4, which represent a *weight* × *state* multiplication, are partly encoded in time (the state variable) and partly amplitude modulated (the weight variable).

As VLSI device dimensions are reduced, the operating voltage is often decreased (a 3.3 V operating voltage is now quite common). This means the range over which the amplitude of any signal may be modulated also decreases. Sometimes it may be possible to redesign the circuit to allow for the scaling, but sometimes it may not. Furthermore, signal-to-noise ratios and dynamic ranges worsen as the process is scaled down. In short, signals that are amplitude modulated scale poorly as the technology is scaled.

This raises a further question: do signals that are time encoded suffer this disadvantage? In principle, the answer is 'no'. In practice, as with everything in the real world, the answer is not so simple. Establishing the exact timing of events (for example, the rise or fall point of a pulse edge) is difficult because of small variations in rise and fall times, and so pulses are susceptible to jitter. Furthermore, the circuits used to produce the pulses (for example the comparator presented in section 4.8), can themselves be scaling dependent.

However, it would be fair to say that encoding signals in time gives additional freedom in the design of analogue circuits. At one extreme, if the designer is creative enough to encode all signals stochastically (using PFM), the circuits can achieve a measure of scaling independence. At the other extreme, if all signals are amplitude modulated, then the circuits are heavily scaling dependent, as device dimensions and operating voltages are reduced. Many of our research group's circuits combine signals that are amplitude modulated with signals using encoding in time, and so these circuits lie somewhere between the extremes. Changes in technologies certainly present many challenges to the analogue designer, and pulse-stream techniques can help in meeting these challenges.

4.12   ANN, and other, applications

So far our research group has adapted the design principles described here for MLPs, for on-chip learning, for RBF (radial-basis-function) networks, for robotics and for analogue filters.

The basic pulse-stream technique can be brought to bear upon other applications where its combination of pseudo-analogue behaviour, robustness and simplicity are valuable. The only fundamental constraint is the level of difficulty of the analogue function to be implemented.

The primary disadvantage of the technique is that effort has to be expended in developing additional analogue and digital circuitry to produce and to store the data encoded in the pulses, and to generate some rather unusual-looking ramps. Even there, we have found that the solutions often reduce to tried-and-tested digital techniques.

4.13   Further reading on applications of the technique

The majority of applications to date are neural ones, but most authors think hard about how they might develop their techniques for uses outside the field.

Meador and Hylander [63] have designed a pulse-coded, winner-take-all network. The network measures the distance between an input vector, representing some pattern that exists in the outside world, and stored 'prototype' vectors, to determine which prototype most nearly approximates the input. This kind of network, common in neural applications such as self-organizing feature maps, can also be useful for vector quantization and coding, and for statistical data clustering.

Some approaches are explicitly biological. De Yong and Fields have used their knowledge of biological signals as inspiration for applications in signal processing and control systems [26]. Biological neurons communicate using trains of pulselike action potentials, fired continuously or in bursts. The pulse-trains encode timing, frequency and phase relationships that the authors exploit in their artificial networks with the aim of storing complex patterns for signal processing problems.

Elias uses fairly simple models of very-low-level neural structures, namely the dendrite (one of the structures of nerve cells to which synapses connect) [27]. His work is interesting on two counts. The first is that his circuits can realize temporal encoding, a well known feature of real neurons; he stimulates chains of simple RC circuits that emit different responses depending on the physical distance of the stimulus from the output. The second matter of interest is that, by using the spatial characteristics of several dendrites working in parallel, he is able to build useful feature detectors that can respond to, for example, lines moving in particular directions.

The three approaches described so far in this section can all be found in an interesting collection of papers in [64], where you can also find several other ideas ranging from the realization of Boolean functions [24] to simple interfacing of networks to the analogue world using pulse-density modulation [95].

Finally, we mention an application from our own group. Papathanasiou and Hamilton, have moved away altogether from neural structures, using pulse-coding for a filter building-block they call the Palmo filter [72]. Fundamental to filter structures is the integrator, which they realize using pulse-stream techniques. Somewhat like the synapse shown in figure 4.4, their design uses PWM to represent the magnitude of signals. The pulses control the ON-time of transistors to gate positive and negative currents that are accumulated on a capacitor as the currents charge and discharge it. They have found ways of reducing the effects of variations in process, and intend to design a 'programmable' filter chip with an array of filter taps that can implement a range of different filters.

Various other aspects of our research group's activities, including some of those described here, are illustrated in [42], in [47], which applies neural techniques to robotics, and in [114] where the technique is used for circuits for on-chip learning.

Chapter 5

Cellular Neural Networks

Mark Joy
Kingston University, UK
ma_s443@crystal.kingston.ac.uk

5.1 Introduction

Evolving physical systems have been extensively studied by mathematicians and scientists through the methods of ordinary differential equations (ODEs) and partial differential equations (PDEs). These approaches differ in their emphasis on state variables: time is the sole independent variable for ODEs, in contrast to PDEs which exhibit time *and* space as independent variables. Models of physical systems may also be differentiated by the nature of the variables we use, whether discrete or continuous. Proceeding in this fashion, we arrive at a classification of physical models. Now while classification is to be greatly desired, we should not forget that we are at liberty to ignore the boundaries between types. Numerical methods for solving PDEs exist which involve a 'discretization' of space and a consequent approximation by a system of ODEs. Similarly, one can study flows by taking snapshots of the evolution, proceeding by discretizing time. On the other hand, we may envisage systems of interacting but distinct cells, whose state variables are continuous; such is the case in Turing's cellular model of morphogenic processes. Now the discreteness in our modelling of the latter system has not been introduced as an approximation, but is implicit in the system architecture. Here lies an interesting point: although modelled by ODEs, which emphasize the time evolution of the cells, we are as much interested in the spatial patterns which may arise. This point of view suggests that we consider the mathematical description a space–time one, in the manner of PDEs, but based on a discrete space. We may proceed further. In recent years mathematical modelling has been extended to include studies of systems which are described in discrete time and on a discrete space—such models are referred to as *cellular automata*, CA, invented by von Neumann and

Ulam [112]. Such systems consist of cells whose activations s_i assume one from a number of discrete state values. Interactions take place among cells in a local neighbourhood; for example, in one dimension a cellular automaton has a dynamic rule of the form:

$$s_i(n+1) = f(s_{i-k}(n), s_{i-k+1}(n), \ldots, s_i(n), \ldots, s_{i+k}(n)) \qquad (5.1)$$

where k is the size of the neighbourhood. Apart from a coarse, qualitative description of different types of CA dynamics (see, for example [113]), little is known. Such systems, though, provide a tentative approach to the problem of providing computational models of physical reality.

So much for discrete systems. In the context of neural networks we often seek to describe how large populations of cells are capable of transporting and processing information by the application of dynamical rules described by ODEs. Various types of collective behaviour in neuron populations, such as cooperation and competition, achieve the stable organization which we witness in the brain. Based on this knowledge, we are now beginning to provide adequate differential equations for biological nets. However, many of these models suppose global interconnection schemes, something that nature seems to avoid. It is well known that neural networks which have evolved to solve modal problems such as vision, are large-scale systems composed of units which possess local connections, or *receptive fields*, through which signals propagate. There is an hierarchical organization in the large-scale systems in the brain. At the lowest level in the hierarchy are the receptor neurons, which react to a specific, finite area (the receptive field) of the sensory space. These receptors converge onto second-order neurons in the central nervous system which also have receptive fields constructed from a synthesis of first-order receptive fields. Typically, this hierarchical organization is propagated through several layers; for example, the retinal organization conforms to such a scheme: the centre-surround retinal and lateral geniculate cells possess receptive fields which are connected to cortical cells of various types. At the highest level of this particular hierarchy, the receptive fields of the cortical cells overlap in complex ways, see [56]. What we are observing here is the mechanism of computation, via local dynamic rules, which we witnessed in CA.

In the next section we intend to pursue this theme by studying a relatively new type of model, one which has a set of dynamical equations based on local neighbourhood connections.

5.2 The CNN paradigm

We now introduce *cellular neural networks*, CNNs, and investigate their elementary properties. CNNs are particularly well suited to solving image processing tasks, since many problems of this type can be solved by phrasing the correct local rules on a cellular system, as we shall see.

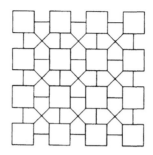

Figure 5.1. A two-dimensional CNN of size 4 by 4.

The architecture of the generic CNN is a two-dimensional grid of cells, seen in figure 5.1.

The individual cells are nonlinear circuits with connections only to their 'closest' neighbours. In this context we define the *1-neighbourhood* of a cell, shown in figure 5.1, to be all cells lying at a king's-walk of length at most one on the chessboard arrangement of cells; thus a 1-neighbourhood consists of nine cells. Now the inner structure of CNN cells, as nonlinear circuits, is unimportant to us. What is important is the set of ODEs which derive from the application of Kirchoff's laws. For the inner cell $C(i, j)$ lying in its 1-neighbourhood $N(i, j)$, we have:

$$\dot{x}_{ij} = - x_{ij} + \sum_{C(k.l) \in N(i.j)} A(i, j; k, l) f(x_{kl})$$

$$+ \sum_{C(k.l) \in N(i.j)} B(i, j; k, l) u_{kl} + I_{ij} \qquad (5.2)$$

where x_{ij} is the state variable associated with this cell. $A(i, j; k, l) f(x_{kl})$ is a current source in cell $C(i, j)$, arising from its connection with cell $C(k, l)$; $f(x_{kl})$ is a nonlinear current source in each cell $C(k, l)$, which we take as its *output*. The characteristic of f is a ramp sigmoid, seen in figure 5.2.

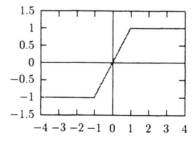

Figure 5.2. The ramp sigmoid.

The term $B(i, j; k, l)u_{kl}$ represents another influence on the inner cell, due to the input, u_{kl}, to cell $C(k, l)$. Finally, I_{ij} represents a clamped input.

Equation (5.2) conforms to the *additive neural network* equation (see [40]), with the first summation term constituting the *feedback* in the network and the second summation term representing *feedfoward* connections between cells. Other generalizations of equation (5.2) exist, but will not concern us here. We refer the interested reader to the review of the CNN model provided by its author [16].

The set of *weights*, $A(i, j; k, l)$ and $B(i, j; k, l)$, can be considered to be space invariant in the sense that they remain the same in all neighbourhoods and depend only on their position relative to the inner cell of any neighbourhood. We collect the $A(i, j; k, l)$ and $B(i, j; k, l)$ together into a pair of templates, the feedback and feedforward templates:

$$
\begin{array}{ccc}
A(-1, -1) & A(-1, 0) & A(-1, 1) \\
A(-1, 0) & A(0, 0) & A(0, 1) \\
A(-1, 1) & A(0, 1) & A(1, 1)
\end{array}
$$

$$
\begin{array}{ccc}
B(-1, -1) & B(-1, 0) & B(-1, 1) \\
B(-1, 0) & B(0, 0) & B(0, 1) \\
B(-1, 1) & B(0, 1) & B(1, 1).
\end{array}
$$

In this way we generate the interconnection scheme of the CNN by translating both templates across the two-dimensional grid of cells.

As a set of nonlinear ODEs, equation (5.2), and its generalizations exhibit dynamical behaviour that would be beyond the scope of this discussion to describe. However, there are still fundamental questions to be answered in this direction:

(i) *What is the most appropriate way of describing stability for CNNs?*

(ii) *What guarantees exist that the CNN circuit will be stable?*

One crude form of stability that we are able to exhibit and which we will use later in this chapter is that of *positive Lagrange* stability. A set of ODEs defined on Euclidean space $\mathbf{R}^n$, is (positively) Lagrange stable if all solutions lie in a bounded subset: this means of course that as $t \to \infty$ no solution $x(t)$ can escape to infinity. Now neural network dynamical systems are always defined on compact subsets of $\mathbf{R}^n$. Consider, for example, the CNN (5.2). It is possible by boundedness of the ramp sigmoid f to choose a sufficiently large ball, centred at the origin, for which $\dot{x}_i < 0$, for all i, on the spherical boundary of the ball (just pick $|x_i|$ greater than the modulus of the other terms on the right-hand side of equation (5.2), then take the radius of the ball to be larger than each of these chosen values, $|x_i|$). Clearly no trajectory can escape such a ball, and for this reason the CNN dynamical system is compact.

We will postpone a more detailed discussion of questions (i) and (ii) until the next section. However, in the context of a CNN as a parallel image processor the following important considerations arise: we assume that we have an input image, consisting of an array of $M \times N$ 'grey-scale' pixels; that is an array of reals lying in the range -1 to $+1$ (-1 here is a completely white pixel, 1 is a black pixel). As an image processor we may require the CNN to extract 'features', edges, corners and so forth, and other tasks may require transformations on the image, for example, shadowing or the detection of moving objects in a presented sequence of images. In all modes of operation we will demand that the CNN settles, in the steady state, to an output image consisting of some d.c. operating point of the circuit, equivalently, an equilibrium $\bar{x}$, of the set of ODEs—thus a CNN is an example of a *point attractor* neural network; more on this in the next section. We then take the output of the CNN as $(f(\bar{x}_1), \ldots, f(\bar{x}_n))$.

To see how all this works, let us see a CNN in action. Our image processing task will be to extract the edges of a square appearing in an input 'bipolar' image. We have said before that many image processing tasks can be coded as a set of local rules to be performed in parallel across the image. Let us see how a heuristic argument can supply us with the local rules in this case.

In the spirit of the discussion in the introduction to this chapter, suppose that we have a large number of pixels in the image, so large in fact that we agree to consider the CNN grid as a continuous medium. State variables are replaced now by a function $u(x, t)$ say, which depends both on spatial position in the medium, x, and time t. In classical settings such as the heat equation, the change in heat at x at time t, is proportional to Δu at x, at that particular moment; here Δ is the *Laplace* operator. Now, as is usual in the calculus, in order to make computations, we may discretize the medium and make an approximation to the PDE. A system of ODEs is revealed having an equation for each discretized spatial location. For example, the two-dimensional heat equation:

$$u_t = \Delta u(x, y, t) \tag{5.3}$$

is transformed into the system of ODEs:

$$\dot{u}_{ij} = -u_{ij} + \tfrac{1}{4}(u_{ij-1} + u_{ij+1} + u_{i-1j} + u_{i+1j}) \tag{5.4}$$

by appropriate approximations of the Newton quotients. We may approach the edge extraction problem in a similar fashion. Consider the edge of a square in profile: the plateaux of pixel colours have a 'colour gradient' u of zero, the sharp transition between them gives a 'spike gradient' as shown in figure 5.3.

In contrast with the classical heat equation, (5.3), which tends to smooth 'wrinkles' in the temperature gradient (in the one-dimensional case, u decreases if u is convex downwards, and increases if convex upwards), here we wish to emphasize such fluctuations so that edge pixels increase with respect to their neighbours. Our feedback template should therefore encode something like

Figure 5.3. (a) Colour edge. (b) Colour gradient.

$u_t = -\Delta u$, in order to emphasize the colour gradient. Our local rule is nearly complete, the last thing left to do is to add a negative (constant) external forcing term, I, to ensure that for cells initially (at time zero) lying in colour plateaux, the pixel value decreases as $t \to \infty$. Such cells will tend to turn white (-1). On the profile transition, or edge of the square, the gradient 'spikes' with a positive value. By a suitable choice of $|I|$ we can ensure that for these cells, the pixel value increases monotonically as $t \to \infty$. In order to extract an output we apply the squashing function f.

The reader may check, by reference to equation (5.4), that we end up with a CNN feedback template, scaled appropriately to:

$$\begin{array}{ccc} 0 & -1 & 0 \\ -1 & 4 & -1 \\ 0 & -1 & 0. \end{array}$$

This feedback template gives us the CNN equation (5.5), below. It is important to notice the following consequence of the piecewise linearity of f: if $|x_{kl}| \leq 1$, then the above equation is simply a discretized version of the linear PDE $u_t = -\Delta u$. However, if a cell x_{kl} *saturates*, that is $|x_{kl}(t)| > 1$, at some time t, then it 'drops out' of (5.5); that is, (5.5) becomes independent of x_{kl}

$$\dot{x}_{ij} = -x_{ij} + 4f(x_{ij})$$
$$- \tfrac{1}{2}(f(x_{ij-1}) + f(x_{ij+1}) + f(x_{i-1j}) + f(x_{i+1j})). \qquad (5.5)$$

Whilst saturated, therefore, cell $C(k, l)$ makes no contribution to the dynamic behaviour of any cell to which it is attached. We should also notice that equation (5.5) is a nonlinear system of ODEs, because f is nonlinear. By contrast, $u_t = -\Delta u$ is a linear PDE. In short, our heuristic argument really is only a way of thinking about this problem.

Now, the template above does indeed extract the edges of a square. Unfortunately, extraneous cells are extracted as well; let us call these cells *rogue* cells, shown in figure 5.4.

Examining the evolution of the system we observe that if a cell lies at the edge of the square its pixel value initially decreases from $+1$, and therefore

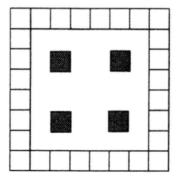

Figure 5.4. The rogue cells (shaded).

unsaturates, so that it may influence the behaviour of cells to which it is connected. However, such cells become relatively larger than neighbouring cells, as $t \to \infty$. Indeed, this is the behaviour we had in mind when constructing this template. Due to the interaction of edge cells though, the rogue cells have transients that initially have negative derivatives, but become positive at a later time. This behaviour causes rogue cells to tend towards $+1$, as $t \to \infty$.

Such problems are typical and are caused by loss of the information of the original image. The solution is to use a feedfoward template and use the initial image as input to each cell, *as well* as the initial condition to the circuit. No information will be lost and we end up with the CNN design:

$$\begin{matrix} 0 & 0 & 0 \\ 0 & 2 & 0 \\ 0 & 0 & 0 \end{matrix}$$

$$\begin{matrix} 0 & -0.25 & 0 \\ -0.25 & 2 & -0.25 \\ 0 & -0.25 & 0 \end{matrix}$$

for a suitable choice of I, for example $I = -1.75$, will do.

Of course, as neural networks, CNNs can learn templates for specific tasks. Many algorithms exist to determine templates, some of which are based on relaxation methods for solving algebraic inequalities. The methods of genetic algorithms have recently been applied to CNNs as well. For lack of space we will not discuss the wider issue of learning in CNNs in this chapter. What has emerged though is an extensive library of templates for a host of different image processing tasks.

This example serves to illustrate the design of a CNN. As an example of an 'averaging method' for solving an image processing task, such an approach is not new, indeed the terminology *cloning template* has been borrowed from two-dimensional digital filter theory. The interest in CNNs lies elsewhere and this is an issue which we now need to address.

The CNN model is simply a special case of the additive neural network model. Its speciality arises from the choice of f_j and the connection, or weight matrix. Although it is usual to define the saturation functions f_j as ramp sigmoids, in practice of course, components of nonlinear circuits, operational amplifiers and so forth, will have characteristics which are smooth $(C^1\text{-})$ approximations to the ramp sigmoid. Therefore, the most important feature of the CNN model is its local connectivity.

The promise of CNNs is the successful fabrication of such a device onto a chip. Currently, fully connected nets are tethered to the digital computer as a piece of software. VLSI implementation is virtually impossible with a massively parallel architecture such as a Hopfield net since the heat generated by the circuit elements cannot be dissipated. Therefore the only way in which a fully connected net can see the light of day is as a computer program. One can circumvent the heat dissipation problem by reducing the number of connections, and this was the central idea when the CNN was first proposed. The fact that the CNN has already found its way onto a chip is some measure of success. The fully tested 6×6 CNN chip reported in [21], boasts approximately 0.3 tera XPS; that is, 0.3×10^{12} analogue operations per second. This is indeed promising, but there is a long way to go. At the moment only 'toy' applications exist for CNNs: we will require chips with many more cells in order to perform useful real-world applications and these seem some way off.

At this point I would like to offer a personal justification for considering neural networks as cellular systems. The brain provides us with a constant reminder of the fundamental nature of local organization in biological computation and it seems to me that we should attempt to reflect this in our models and equations. Generalizations of the linear CNN model which we consider in this chapter provide for templates which may be space variant, nevertheless, all CNN models are locally connected. For this reason I feel that CNNs offer important examples of neural network dynamical systems, examples that reflect the way in which nature has fashioned some of the most important biological nets. We turn to the study of such systems in the next section.

5.3 CNN dynamical systems

In recent years an explosion of interest has developed in nonlinear dynamical behaviour. This interest is reflected in a growth of articles, both in the popular scientific press and in research journals, describing the more exotic examples, including chaotic dynamical systems. At the same time, it has been conjectured by neurobiologists that chaos may have an important role to play in biological dynamics. For example, in the olfactory system of the rabbit we may 'lump' together cells in the bulb and consider the resulting dynamical system; what is interesting is that we witness chaotic dynamics in the absence of learned odours. Perhaps more interesting though, is the following mechanism: upon inhalation of a previously learned odour the dynamics changes in an essential way (bifurcates), and the state of the system is attracted to a periodic evolution. The latter behaviour corresponds to and represents the learned odour—we see no search through a memory store. For more details, we refer the interested reader to [88].

This description of neural dynamics is controversial and has attracted much criticism, primarily because of its insistence upon the necessity of chaos for learning in biological nets, but what it does illustrate is the use of dynamical systems theory. Chaotic dynamics, as a mathematical theory, is in its infancy and, despite its popularity, offers little in the way of good, predictive theorems. We therefore find it more profitable to turn our attention to dynamical behaviour which we find easier to interpret and which will satisfy the more modest aims which we have in mind here, namely, a study of neural dynamics from the perspective of local interactions, and the usefulness of the CNN model in this study.

We have seen the suggestion of chaotic dynamics as a driving mechanism for learning in neural networks, but what other types of behaviour can be useful? Most nets which have been studied have exhibited *convergent* dynamics. In order to define such behaviour let us briefly introduce some other definitions which we will find useful, and which we will exploit throughout this section.

An ODE

$$\dot{x} = f(x) \tag{5.6}$$

where $f: U \to \mathbf{R}^n$, a *Lipschitz* (see below) function defined on an open subset U of $\mathbf{R}^n$, defines a *dynamical system* in the following way: we collect together the solutions or *trajectories*, $\phi(x, t)$ to (5.6), and define a *dynamical system* to be the collection of mappings $\phi_t: \mathbf{R}^n \to \mathbf{R}^n$, which take a point x to its position at time t, $\phi(x, t)$. If U is a bounded set in $\mathbf{R}^n$, we may take the collection of maps to be defined on all $t \in \mathbf{R}$. Intuitively, these maps exhibit the time-t evolution of the system.

The use of a Lipschitz function is a technicality which gives us *unique* solutions through initial conditions, and we require unique evolution through an initial point, in order for the definition of a dynamical system to make sense. In fact a C^1-smooth function will do, but ramp sigmoids are not even differentiable at the corners, so we use Lipschitz functions: functions f, which satisfy $|f(x) - f(y)| < \kappa |x - y|$, for all $x, y \in U$, where $\kappa > 0$.

We are now able to make the definition which prompted these technicalities: a dynamical system is said to be *convergent* if every trajectory $\phi(x, t)$ converges to an *equilibrium* solution as $t \to \infty$. An equilibrium point is a rest point of the system—it sits still as the system evolves, that is, it undergoes no change under the time-t evolution maps; for example, the downward resting point of a pendulum is an equilibrium. It may be checked that equilibria are zeros of f.

The reason that we require our nets to be convergent is connected with the representation of information. We are all accustomed to storing data and numbers, in particular, as discrete entities, for example as bytes of memory in digital computers. If we think of neural systems as retrieving information, it is natural to require this information to be in the form of a constant collection, say of n real numbers. If a nonlinear circuit has recovered or created this information, such a point is a d.c. operating point of the circuit. In terms of dynamical systems the required concept is clearly that of an equilibrium point. A convergent neural network dynamical system is an example of what we have called a point attractor network. Other authors, particularly Hopfield, have coined the phrase *content addressable memory* or CAM. The CAM algorithm is pure analogue: we represent partial or corrupted versions of the network memories as points $x \in U$ and follow the evolution of the point towards an equilibrium $\bar{x}$, which represents the recovered memory.

Since the CNN model is of the additive type, what we seek are theorems which guarantee the convergence of additive neural networks. Firstly, let us review the better known results on the convergence of neural nets and then consider how CNNs, by providing examples worthy of close study, lead us towards new and interesting convergence theorems.

The first important result in this direction is the celebrated *Cohen–Grossberg theorem* [18]. This theorem applies to a wide range of neural models, including fully connected Hopfield networks, as well as CNNs. Provided that the connection scheme is symmetric and that some other mild conditions are met the theorem states, for example, that a neural network dynamical system possesses a *strict Liapunov* function V. The existence of a strict Liapunov function is one of the commonest ways of establishing convergence for dynamical systems. Such a function is a scalar function V, which decreases strictly along trajectories, that is: $\mathrm{d}/\mathrm{d}t\, V\,(\phi(x, t)) < 0$. Roughly speaking, if we embed our system in an Euclidean space of one higher dimension, using the graph of V, then trajectories follow paths on this high-dimensional surface which always lose height (V); eventually trajectories settle in 'hollows' on this surface. We may think of V as being a 'generalized' energy which is dissipated as the system evolves, so that trajectories settle at points of minimum energy. Readers who are familiar with multilayer perceptrons and the backpropagation algorithm as a gradient descent method, can think of Liapunov functions as a continuous-time version of such a method, except that trajectories are not stochastic.

Cohen and Grossberg originally studied competitive systems of the form:

$$\dot{x}_i = a_i(x)\left(b_i(x_i) - \sum_k c_{ik} d_k(x_k)\right) \tag{5.7}$$

where (c_{ik}) is a symmetric matrix, $a_i \geq 0$ and $d_k' \geq 0$, so that we may take a sigmoid function for d_k. Thus we may view (5.7) as an additive neural network with symmetric connections, together with an amplification function a_i (we can of course take $a_i \equiv 1$), and self-signalling functions b_i. The Liapunov function discovered by Cohen and Grossberg is:

$$V(x) = -\sum_i \int_0^{x_i} b_i(\xi) d_i'(\xi)\,d\xi + \frac{1}{2}\sum_{jk} c_{jk} d_j(x_j) d_k(x_k). \qquad (5.8)$$

The reader may wish to verify the following inequality:

$$\frac{d}{dt} V(x(t)) = -\sum_k a_k(x) d_k'(x_k) \left(d_k(x_k) - \sum_j c_{kj} d_j(x_j) \right)^2 \le 0 \qquad (5.9)$$

establishing the basic requirement of any Liapunov function. Hopfield [45] rediscovered essentially the same function for the special case of (5.7), the Hopfield network:

$$\dot{x}_i = -\mu_i x_i + \sum_k T_{ik} g(x_k) \qquad (5.10)$$

where (T_{ik}) is symmetric and in this case, $T_{ii} = 0$. One simply substitutes $a_i = 1$, $b_i = \mu_i x_i$ and $T_{ij} = -c_{ij}$, into (5.10) to obtain Hopfield's energy function:

$$\sum_i \mu_i \int_0^{x_i} \xi g'(\xi)\,d\xi - \frac{1}{2}\sum_{jk} T_{jk} g(x_j) g(x_k). \qquad (5.11)$$

We may also use (5.10) to provide us with a Liapunov function for symmetric CNNs. Although the theorem insists upon regularity conditions on the activation functions which are not present if we use ramp sigmoids, we may still prove by a direct formal analogy that:

$$V(x) = -\frac{1}{2}\sum_{ij}\sum_{kl} A(i,j;k,l) f(x_{ij}) f(x_{kl}) + \frac{1}{2}\sum_{ij} f(x_{ij})^2$$

$$+ \frac{1}{2}\sum_{ij}\sum_{kl} B(i,j;k,l) f(x_{ij}) u_{kl} - \sum_{ij} I_{ij} f(x_{ij})$$

is a Liapunov function for a symmetric CNN with feedback and control templates A and B respectively, constant inputs u_{kl}, and clamped inputs I_{ij}. Notice that we have provided an answer to question (ii) from section 5.2: the symmetry of the connection weights is a sufficient condition to guarantee convergence (stability) of the CNN. Notice also that the CNN equations are doubly indexed rather than the usual, singly indexed additive neural network equation; taking this into account, we have substituted $a_i = 1$, $b_i(x_i) = -x_{ij} + \sum_{kl} B(i,j;k,l) u_{kl} + I_{ij}$ and $c_{ij} = A(i,j;k,l)$ into (5.10) to obtain the Liapunov function. The Cohen–Grossberg Liapunov function is remarkably versatile, primarily because so many systems can be described by the ODE (5.7)—from interacting species consisting of predators and prey, to shunting cooperative–competitive recurrent networks.

What theorems do we have if the connection matrix is not symmetric? Recently, work by Hirsch [44] has resulted in a series of striking results which we may apply directly to the convergence problem for CAM. The most important result concerns cooperative, irreducible dynamical systems:

(i)   Given an ODE $\dot{x} = f(x)$, $x \in U$, the resulting dynamical system is called *cooperative* if $\partial f_i(x)/\partial x_j \geq 0$, for all $x \in U$.

(ii) An ODE is *irreducible*, if for all $1 \leq i \neq j \leq n$, there exists a chain of indices $i = k_0, k_1, \ldots, k_m = j$ such that:

$$\frac{\partial f_{k_{r-1}}}{\partial f_{k_r}} \neq 0.$$

In the neural network context, irreducibility means that the output of any cell can indirectly influence the activations of all other cells, and is a mild condition to impose on dynamical systems. Hirsch's important result [44] is:

Theorem. Given a cooperative, irreducible dynamical system, *almost* every trajectory approaches the equilibrium set $\mathcal{E}$. Hence if $\mathcal{E}$ is finite, *almost* every trajectory converges to an equilibrium point.

Almost, here, is in the sense of Lebesgue measure: what the theorem says is that there may well be trajectories which do not converge to an equilibrium, but $Q = \{x \in U \mid \phi(t, x) \text{ does not converge}\}$, will be a subset of Lebesgue measure zero. For readers who are unfamiliar with measure theory, a subset Q of measure zero is 'thin' and 'insubstantial', like an algebraic curve in a plane, or a plane in $\mathbf{R}^3$; if $x \in Q$, then any small perturbation of x in almost every direction 'kicks' x out of Q. In this case therefore, non-convergence is seen as *exceptional* behaviour: if x is a point whose evolution does not end at an equilibrium, there are arbitrarily close neighbours having such an evolution, and x is an exception rather than the rule. Another way of stating this is to say that with probability one, a point x has a convergent future.

Now Hirsch's theorem is directly applicable to CNNs provided we utilize an A-template whose entries are nonnegative. Firstly though, consider the structure of the equilibrium set $\mathcal{E}$. Suppose equilibria are isolated; that is, if $\bar{x}$ and $\bar{y}$ are two equilibria, then $|\bar{x} - \bar{y}| > \delta$, for some $\delta > 0$. In this case we can show that there are a finite number of equilibria, $\mathcal{E} = \{\bar{x}_1, \ldots, \bar{x}_n\}$, say.[1] As a trajectory approaches the equilibrium set, in accordance with Hirsch's theorem, it must actually approach a *particular* point. Thus, isolated equilibria imply an almost convergent dynamical system. Although we will not investigate such matters here, even if a neural network is represented in some set of coordinates by a non-cooperative ODE, there may actually be a set of coordinates in which the network is cooperative. Therefore the 'natural' activation coordinates are not always the most illuminating to deal with, and may obscure important clues as to global dynamic behaviour.

So far, our discussion has been broad enough to embrace both the general, additive neural network category and the CNN category of dynamical systems.

---

[1] This follows from the compactness of the phase space of the dynamical system, as any reader who is familiar with point-set topology will appreciate.

We now outline some convergence theorems solely for CNNs. The hypotheses of these results seem quite natural when imposed on CNNs, but seem too restrictive for fully connected networks. In other words, the hypotheses seem unlikely to be satisfied in the fully connected case—there are simply too many weights or connection strengths to take care of. Now convergence theorems for nonlinear differential equations are hard to come by; from this point of view, the forthcoming results are of interest in themselves. For neural network researchers, such theorems are also to be desired for we are constantly seeking general principles or conditions which imply emergent behaviour of some sort or another. We may observe that on the one hand a fully connected net will not, in general, satisfy the conditions of these theorems, and on the other hand such results enrich our knowledge concerning what types of ODEs can provide us with prototype CAM. It seems both reasonable and profitable therefore, to consider neural networks with local connection schemes, exemplified by CNNs.

I now wish to present a stability theorem to give the flavour of these results, indicating what it is possible to prove in this category. For the moment then we consider a CNN, described by the dynamical equation:

$$\dot{x}_i = -x_i + Af(\boldsymbol{x}) \tag{5.12}$$

where $f(\boldsymbol{x}) = (f(x_1), \ldots, f(x_n))$, is the Cartesian product of the ramp sigmoid function f, and A is a sparse matrix exhibiting the special structure of a CNN feedback matrix, obtained from the template T_A, say. Our task is to describe the dynamics of (5.12). We certainly know the dynamics of a single, isolated cell, since it is governed by a scalar differential equation:

$$\dot{u} = -u + pf(u) \tag{5.13}$$

for some real $p > 1$. (5.13) is easy to analyse: the reader may verify that the system is convergent with stable, attracting equilibria at $\pm(p-1)$ and an unstable equilibrium at 0. We require the same sort of behaviour for the system as a whole; that is, we require *every* cell to saturate. Here we introduce an assumption which may be realistic for CNNs but too restrictive, if not, unrealistic for fully connected nets. Let us assume that the self-feedback term, $A(0,0)$, in the template T_A, dominates in the following sense:

$$A(0,0) - 1 > \sum_{jk} |A(j,k)|. \tag{5.14}$$

What we are essentially saying here is that the self-feedback of each cell dominates over the influences of the cells to which it is attached. This follows by examining the Jacobian matrix $A - I$ in (5.12). We find it is *row diagonally dominant*; the diagonal element is greater then the moduli of the other elements in that row:

$$A_{ii} - 1 > \sum_{ik} |A_{ik}|. \tag{5.15}$$

The right-hand side of (5.15) represents the maximum possible net input, contributed by its neighbours, to the cell with trace x_i. Clearly, self-feedback is the dominant signal here. Now in this case it is possible to prove that the dynamical system associated with (5.13) is convergent. We now sketch the proof; the reader may skip the following paragraph if he or she wishes.

By piecewise linearity of the ramp sigmoid, there are subregions of the *phase space*—the set on which the dynamical system is defined—on which the restricted system is linear. These regions are defined by the sets of saturated and unsaturated state variables. In any linear subregion, where there is at least one unsaturated state variable, there is at least one eigenvalue of the Jacobian matrix with positive real part, again this follows from the dominance hypothesis (5.14). Now look at the formula for the solution to the linear differential equations in these subregions: $x(t) = x_0 \exp t J$, where J is the Jacobian matrix. Clearly an eigenvalue with a positive real part implies that $x(t)$ is expanding in some direction. Therefore no trajectory can remain in these linear subregions. Now in order to leave such a region a trajectory must cross one of the hyperplanes defined by $x_i = \pm 1$. It is here that the dominance hypothesis has its final pay-off: along these hyperplanes the vector field points 'outwards', that is in the direction of increasing $|x_i|$, as shown in figure 5.5.

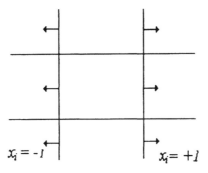

Figure 5.5. The direction of the vector field along the hyperplanes.

The consequence of this geometrical configuration is that once crossed, trajectories can never recross such a boundary. Essentially this means that all cells will ultimately saturate. For full details of such an argument see [49].

Notice that the dominance condition is a fairly natural hypothesis to place on a large-scale system, that is, a system comprising of many subunits. What it means is that the connections between subunits are so weak that each subunit behaves as if isolated. In a fully connected net though it is unlikely that such a condition would hold; in a CNN it can be imposed easily, through the mechanism of local connections.

Incidentally, another important kind of stability holds for a CNN satisfying (5.14): we are able to show that such a dynamical system is *structurally stable*, as well. The notion of structural stability dates back to the Russian school of applied mathematics in the 1930s which investigated dynamics that were stable under perturbations of the equations defining the system *itself*. We refer the reader to [3], or any modern text on dynamical systems for precise definitions of this kind of stability.

What structural stability requires is that under sufficiently small changes in the system equations, qualitatively the 'same' dynamics will persist. The philosophy here is that since one cannot know, with infinite precision, the value of the parameters associated with the system equations—we cannot measure nature exactly—our model should capture the essential dynamical features despite inaccuracies in our measurements. Therefore mathematical models with very nearly the same parameters, if structurally stable, should exhibit the same important features. Now work by Palis and Smale in the late 1960s [71] gave sufficient conditions for a dynamical system on a compact space to be structurally stable. Basically, these conditions require that there are no 'saddle connections'; that is, there are no trajectories connecting an unstable equilibrium of the saddle type with another of the same type.

If we now consider a CNN satisfying the dominance hypothesis, the only way in which saddle connections can exist is for trajectories to cross and recross the hyperplanes $x_i = \pm 1$, and we have already seen that this is impossible. It follows from Palis and Smale's work that the CNN dynamical system is structurally stable. This is a very pleasant corollary of a natural hypothesis on CNNs and is a rare example of a structural stability theorem for neural network dynamical systems.

There are many more stability theorems for CNNs and all have hypotheses which can be framed on the template parameters. For lack of space we will not examine these results here, but what I hope the reader has gained from this section is an appreciation of the fact that by making local neighbourhood connections, we are furnished with a category of neural networks for which 'natural' stability theorems may exist in some abundance.

5.4   Conclusions: the CNN as supercomputer or neural network?

In this brief introduction to the subject of CNNs I have offered a personal viewpoint—that of the locally connected neural network dynamical system. Other workers in this field may be uneasy with this emphasis and would not share my position wholeheartedly. For example, some see the CNN model chiefly as an embryonic supercomputer, or at any rate a paradigm for analogue, parallel processing using local rules. Indeed, for some workers in this field this is the underlying rationale of CNNs. If fabrication of these devices proves successful in the near future, then such machines may soon be with us. They would consist of an upper layer of logic, supporting an operating system, connected to a lower layer of analogue processors, capable of implementing image processing tasks as we have described in section 5.2. A name for such machines has even been proposed: *analogic supercomputers*. As more *analogic programs* are written, geometric processing of all kinds could, in theory, be implemented. The technological problems associated with fabricating CNNs are proving difficult to surmount at present, so that our enthusiasm for such devices must be tempered with a touch of realism for now. The interested reader should consult [79], for a discussion of the CNN as supercomputer.

Another promising application area for CNNs seems to be in neurobiological modelling. Importantly, this use of CNNs emphasizes their neural network inheritance. Quite staggering levels of parallelism in the brain support complex visual processing, for example, which is currently beyond the capability of digital computers, furthermore the challenge of describing these biological systems is daunting. Nevertheless, work is beginning: the PSYCHE project—see Chapter 7 of [1]—represents an attempt to model a brain in-the-large. Similarly, researchers have demonstrated CNNs as a versatile tool in the design of artificial retina by showing that the morphological features of its biological counterpart have relatively easy implementations as CNNs. Indeed, so successful has this design proved that most of the 'classical' illusions and motion related effects, which are the artefacts of biological interactions, have been demonstrated in the CNN retina: the Herring grid and the Muller–Lyer illusions, the phi phenomenon and comet-like afterimages (the so-called ' texton theory') are all clearly translated.

Whether we are primarily concerned with VLSI implementation of neural networks in order to solve technological problems, or whether our interest in such systems stems from a search for a fundamental set of properties which have predictive power in terms of neurobiological data, we need to consider seriously the paradigm of local, parallel computation, provided by the CNN model. The future of CNNs as an analogue computing machine rests in the balance; the future for CNNs as an exemplar of a general design principle seems a little more certain, since the wealth of good theorems on the global dynamics of CNNs is adequate enough to justify continued interest from mathematicians and ultimately, engineers. In the words of Stephen Grossberg [40]:

> the set of properties that differentiates one model from another ultimately provides the computational rationale for continued interest in that model, whether because its set of properties better explains an important behavioural or neural data base, or because these properties enable the model to more efficiently solve an important technological problem.

It is the cellular property of local-connectedness that provides the computational rationale for CNNs, and this is a property exploited by nature herself. Can such mechanisms come more highly recommended?

Chapter 6

Efficient Training of Feed-Forward Neural Networks

Martin Moller
Aarhus University, Denmark
mmoller@daimi.aau.dk

6.1 Introduction

Learning in neural networks can be formulated in terms of the minimization of an error function E. The problem of minimizing multivariate, continuous, differentiable functions is one which has been widely studied, and some of the conventional approaches to this problem are directly or with small modifications applicable to the training of neural networks. However, training neural networks differs from conventional optimization problems on two important points:

- The number of variables to be optimized is usually several *magnitudes* larger than the number of variables the conventional optimization algorithms were designed to optimize.
- The derivative information needed to perform the optimization in each iterative step is computationally heavy and is calculated based on very special purpose algorithms.

These two aspects form the major challenges concerned with the design of efficient training algorithms for neural networks.

This chapter reviews different methods for training feed-forward neural networks. The viewpoint is that of optimization which allows us to use results from the optimization literature. These results give information about computational complexity, convergence rates and safety procedures to ensure convergence and avoid numerical instabilities. Throughout the chapter we use results from the PhD thesis by Moller [186] and a paper about first- and second-order optimization methods written by Battiti [123].

The presentation will focus on methods which are especially well suited for training of feed-forward networks. Factors that are important in this classification are computational complexity and the number of problem dependent parameters. In the description of the different methods we make a distinction between *first-order* and *second-order* methods, i.e. between methods based on a linear model and methods based on a quadratic model of the error function.

We emphasize that this chapter is not a full survey of all existing techniques to train feed-forward networks, but a presentation of material judged to be representative for the state of the art within the field. The chapter focuses especially on the scaled conjugate gradient algorithm and its variants.

6.2 Notation and basic definitions

The networks we consider are multilayered feed-forward neural networks with arbitrary connectivity. The network $\aleph$ consist of nodes n_m^l arranged in layers $l = 0, \ldots, L$. The number of nodes in a layer l is denoted N_l. In order to be able to handle the arbitrary connectivity we define for each node n_m^l a set of *source nodes* and a set of *target nodes*.

$$S_m^l = \left\{ n_s^r \in \aleph \mid n_s^r \text{ connects to } n_m^l, r < l, 1 \le s \le N_r \right\} \tag{6.1}$$
$$T_m^l = \left\{ n_s^r \in \aleph \mid n_m^l \text{ connects to } n_s^r, r > l, 1 \le s \le N_r \right\}.$$

The training set associated with network $\aleph$ is

$$\left\{ (u_{ps}^0, s = 1, \ldots, N_0, t_{pj}, j = 1, \ldots, N_L), p = 1, \ldots, P \right\}. \tag{6.2}$$

The output from a node n_m^l when a pattern p is propagated through the network is

$$u_{pm}^l = f(v_{pm}^l), \text{ where } v_{pm}^l = \sum_{n_s^r \in S_m^l} w_{ms}^{lr} u_{ps}^r + w_m^l \tag{6.3}$$

and w_{ms}^{lr} is the weight from node n_s^r to node n_m^l. w_m^l is the usual *bias* of node n_m^l. $f(v_{pm}^l)$ is an appropriate activation function, such as the hyperbolic tangent function. The net input v_{pm}^l is chosen to be the usual weighted linear summation of inputs. The calculations to be made could, however, easily be extended to other definitions of v_{pm}^l. Let an error function $E(w)$ be

$$E(w) = \sum_{p=1}^{P} E_p(u_{p1}^L, \ldots, u_{pN_L}^L, t_{p1}, \ldots, t_{pN_L}) \tag{6.4}$$

where w is a vector containing all weights and biases in the network, and E_p is some appropriate error measure associated with pattern p from the training set. Coordinates of vectors and matrices will, depending on the context, also be referred to by the simpler notation $[w]_i$ and $[A]_{ij}$.

The gradient vector $E'(w)$ of an error function $E(w)$ is an $N \times 1$ vector given by

$$E'(w) = \sum_{p=1}^{P} \left(\frac{\partial E_p}{\partial [w]_1}, \ldots, \frac{\partial E_p}{\partial [w]_N} \right)^T. \tag{6.5}$$

The Hessian matrix $E''(w)$ of an error function $E(w)$ is

$$E''(w) = \sum_{p=1}^{P} \begin{pmatrix} \frac{\partial^2 E_p}{\partial [w]_1^2} & \cdots & \frac{\partial^2 E_p}{\partial [w]_1 \partial [w]_N} \\ \vdots & \ddots & \vdots \\ \frac{\partial^2 E_p}{\partial [w]_1 \partial [w]_N} & \cdots & \frac{\partial^2 E_p}{\partial [w]_N^2} \end{pmatrix}. \tag{6.6}$$

The Taylor expansion of a continuous, differentiable function $E(w)$ is

$$E(w + \Delta w) = E(w) + \Delta w^T E'(w)$$
$$+ \Delta w^T E''(w) \Delta w + \cdots. \tag{6.7}$$

A set $p_1, p_2, \ldots, p_N$ of vectors is said to be mutually conjugate with respect to a matrix A if

$$p_i^T A p_j = 0 \quad \text{when} \quad i \neq j. \tag{6.8}$$

The condition number κ of a matrix A is

$$\kappa = \left| \frac{\lambda_{max}}{\lambda_{min}} \right| \tag{6.9}$$

where λ_{max} and λ_{min} are the largest and smallest eigenvalues of A respectively.

6.3 Optimization strategy

Most of the optimization methods used to minimize functions are based on the same strategy. The minimization is a local iterative process in which an approximation to the function in a neighbourhood of the current point in weight space is minimized. The approximation is often given by a first- or second-order Taylor expansion of the function. The idea of the strategy is illustrated in the pseudo-algorithm presented below, which minimizes the error function $E(w)$ [146]:

1.   Choose initial weight vector w_1 and set $k = 1$.
2.   Determine a search direction p_k and a step size α_k so that $E(w_k + \alpha_k p_k) < E(w_k)$.
3.   Update vector: $w_{k+1} = w_k + \alpha_k p_k$.

4.   If $E'(w_k) -- \neq 0$ then set $k = k + 1$ and go to 2
     else return w_{k+1} as the desired minimum.

Determining the next current point in this iterative process involves two independent steps. Firstly a *search direction* has to be determined, i.e. in what direction in weight space do we want to go in the search for a new current point? Once the search direction has been found we have to decide how far to go in the specified search direction, i.e. a *step size* has to be determined. The accuracy combined with the speed of which an algorithm is able to determine the search direction and the step size determines its overall efficiency. As shall be made clear in the coming sections, there is an important tradeoff between the advantage of obtaining a high accuracy of the search direction and the step size and the time an algorithm can afford to spend on these estimations.

6.4   Gradient descent

Gradient descent is one of the oldest optimization methods known. The use of the method as a basis for multivariate function minimization dates back to Cauchy in 1847 [132], and has been the subject of intense analysis. Gradient descent is based on a linear approximation of the error function given by

$$E(w + \Delta w) \approx E(w) + \Delta w^T E'(w). \tag{6.10}$$

The weight update is

$$\Delta w = -\eta E'(w) \qquad \eta > 0. \tag{6.11}$$

The step size or learning rate η can be determined by a line search method but is usually set to a small constant. In the latter case the algorithm is, however, not guaranteed to converge. If η is chosen optimally in each step the method is often called the *steepest descent method*. The method can be used in *off-line* or *on-line* mode. The off-line mode is the one presented in (6.11), where the gradient vector is an accummulation of partial gradient vectors, one for each pattern in the training set. In the on-line mode, gradient descent is performed successively on each partial error function associated with one given pattern in the training set. The update formula is then given by

$$\Delta w = -\eta E_p'(w) \qquad \eta > 0 \tag{6.12}$$

where $E_p'(w)$ is the error gradient associated with pattern p. If η tends to zero over time, the movement in weight space during one *epoch*[1] will be similar to the one obtained with one off-line update. However, in general the learning rate has to be large to accelerate convergence, so that the paths in weight space of the two methods differ. The on-line method is often preferable to the off-line

---

[1] An epoch is equal to one full presentation of all patterns in the training set.

method when the training set is large and contains redundant information. This is especially true on problems where the targets only have to be approximated, such as classification problems. For further discussion about issues concerning on-line and off-line techniques see section 6.7.

6.4.1 Back-propagation

Until only recently gradient descent was only applicable to single-layer feedforward networks, because a method for the calculation of the gradient $E'(w)$ for multi-layer networks did not exist before that time. A method to calculate the gradient in general was derived independently several times, by Bryson and Ho [128], Werbos [103], Parker [195] and Rumelhart *et al* [80]. The method is now known as the *back-propagation method*, and is central to much current work on learning in neural networks. There is some confusion about what the name *back-propagation method* actually refers to in the literature. Some researchers connect the name to the calculation of the gradient $E'(w)$, others use the name to refer to the gradient descent algorithm itself. We use the name to refer to the gradient calculation. The following lemma summarizes the back-propagation method.

Lemma 6.1. The gradient $E'_p(w)$ of one particular pattern p can be calculated by one forward and one backward propagation. The forward propagation formula is:

- $u^l_{pm} = f(v^l_{pm})$ where
- $v^l_{pm} = \sum_{n^r_s \in S^l_m} w^{lr}_{ms} u^r_{ps} + w^l_m$.

The backward propagation formula is:

- $[E'_p(w)]^{lh}_{mi} = \delta^l_{pm} u^h_{pi}$ $[E'_p(w)]^l_m = \delta^l_{pm}$

where δ^l_{pm} is given recursively by:

- $\delta^l_{pm} = f'(v^l_{pm}) \sum_{n^r_s \in T^l_m} w^{rl}_{sm} \delta^r_{ps}$ $l < L$

- $\delta^L_{pj} = \partial E_p / \partial v^L_{pj}$ $1 \leq j \leq N_L$.

The gradient of the error corresponding to the whole training set is of course a sum of the partial gradients calculated in lemma 6.1. Lemma 6.1 can easily be derived using the chain rule backwards in the network. The main idea of propagating error information back through the network can be extended to the calculation of other important error information features such as second-order information. See [197] and [188] for further details.

6.4.2 Convergence rate

We now turn to the rate of convergence of the gradient descent algorithm. Considering that the negative gradient is the direction of fastest decrease in error, we would intuitively expect to get a fast convergence, if the learning rate η is chosen optimally. This is, however, not the case. If we assume that the error is locally quadratic, then the contours of $E(w)$, given by $E(w) = c$ are N-dimensional ellipsoids with the minimum of the quadratic as the centre. The axes of the ellipsoids are in the direction of the N mutually orthogonal eigenvectors of the Hessian and the length of the axes are equal to the inverse of the corresponding eigenvalues [178]. The gradient $E'(w)$ at a point w on the ellipsoid is perpendicular to the tangent plane in w. This means that the gradient descent direction $-E'(w)$ will not in general point directly to the minimum of the quadratic (the centre of the ellipsoid). The search directions chosen tend to interfere so that a minimization in a current direction can ruin past minimizations in other directions. In fact, the minimization is done in a kind of 'zig-zagging' way, where the current search direction is perpendicular to the last search direction. This is easily verified by the following. Let $d_k = -E'(w_k)$ be the direction of search in the kth iteration. If we minimize in the direction of d_k with respect to η, then

$$\frac{\mathrm{d}}{\mathrm{d}\eta} E(w_k + \eta d_k) = d_k^T E'(w_k + \eta d_k) = 0 \qquad (6.13)$$

$$\Rightarrow d_k^T d_{k+1} = 0.$$

Figure 6.1 illustrates the whole situation. As we shall see in section 6.5, one of the major advantages with conjugate gradient methods is that they approximate non-interfering directions of search.

If we assume that the error function is quadratic with constant and positive definite Hessian,[2] then it is possible to show that the condition number of the Hessian matrix has a major impact on the convergence rate. We can write the error in a neighbourhood of a point w as

$$E(w_k) = E(w) + (w_k - w)^T E'(w)$$
$$+ \tfrac{1}{2}(w_k - w)^T E''(w)(w_k - w). \qquad (6.14)$$

If w_* is the minimum of the error, then $E'(w) = -E''(w)(w_* - w)$ and $E(w_k)$ can be written in the alternative form

$$E(w_k) = \tfrac{1}{2}(w_k - w_*)^T E''(w)(w_k - w_* + E(w)$$
$$- \tfrac{1}{2}(w_* - w)^T E''(w)(w_* - w). \qquad (6.15)$$

The last two terms of (6.15) are constants and can be ignored when

---

[2] This is, however, not always the case.

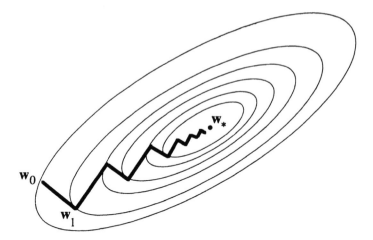

Figure 6.1. The steepest descent trajectory on a simple quadratic surface. Notice that the search directions are perpendicular to each other and to the tangent planes of the contours.

minimizing $E(w_k)$. If the learning rate η is chosen optimal then

$$\eta = \frac{E'(w_k)^T E'(w_k)}{E'(w_k)^T E''(w) E'(w_k)}. \tag{6.16}$$

In order to obtain a bound of the convergence rate we need the following lemma.

Lemma 6.2. The Kantorovitch–Bergstrom inequality states

$$\frac{(x^T A x)(x^T A^{-1} x)}{(x^T x)^2} \le \frac{(\lambda_{max} + \lambda_{min})^2}{4 \lambda_{max} \lambda_{min}}$$

where A is a symmetric positive definite matrix and λ_{max} and λ_{min} are the largest and smallest eigenvalue respectively.

Proof. See [121] or [178].

Based on lemma 6.2 we can now make the connection between convergence rate and condition number.

Lemma 6.3. Assume that the error function is quadratic with constant Hessian $E''(w)$. At every step in the gradient descent algorithm it holds that

$$\frac{E(w_{k+1}) - E(w_*)}{E(w_k) - E(w_*)} \le \left(\frac{\kappa - 1}{\kappa + 1}\right)^2$$

where κ is the condition number of $E''(w)$.

Lemma 6.3 states that the gradient descent method converges linearly with a ratio not greater than $((\kappa - 1)(\kappa + 1))^2$. It can be shown that if the condition number κ is high, then the method is very likely to converge at a rate close to the bound [119]. So the bigger the difference between the largest and smallest eigenvalue, the slower the convergence of the gradient descent method. Geometrically this means that the more the contours of $E(w_k)$ are skewed the slower the convergence of gradient descent. Even if only one eigenvalue is large and all others are of equal size, the convergence will be slow.

6.4.3 Gradient descent with momentum

In [198] the gradient descent update is changed to

$$\Delta w_{k+1} = -\eta E'(w_k) + \alpha \Delta w_k \qquad \eta > 0 \qquad 0 < \alpha < 1 \qquad (6.17)$$

where α is the so called *momentum* term. The addition of this momentum term incorporates second-order information into the method since current as well as past gradient information is taken into account. The change was introduced to avoid oscillations in narrow steep regions of weight space and to increase convergence in flat regions. In [100] an analysis of the effect of the momentum term was given. We briefly summarize these results. The weight update formula given by (6.17) is a special version of a first-order difference equation [104], [3] the solution of which is given by

$$\Delta w_{k+1} = \alpha^k \Delta w_1 - \eta \sum_{i=1}^{k} \alpha^{k-i} E'(w_i). \qquad (6.18)$$

Written in terms of the weights this becomes

$$w_{k+1} = w_k - \eta \sum_{i=0}^{k} \alpha^{k-i} E'(w_i). \qquad (6.19)$$

Equation (6.19) is a first-order difference equation in w with solution

$$w_{k+1} = w_0 - \eta \sum_{j=0}^{k} \sum_{i=0}^{k} \alpha^{j-i} E'(w_i) = w_0 - \eta \sum_{j=0}^{k} \alpha^j \sum_{i=0}^{k-j} E'(w_i). \qquad (6.20)$$

In flat regions in weight space the gradient can be approximated with a constant, say E'. Under this assumption we have

$$w_{k+1} = w_0 - \eta E' \sum_{j=0}^{k} (k - j + 1)\alpha^j. \qquad (6.21)$$

---

[3] The solution for the general equation $x_{k+1} = a_k x_k + b_k$ is: $x_{k+1} = \prod_{j=1}^{k} a_j x_1 + \sum_{i=1}^{k} \prod_{j=i+1}^{k} a_j b_i$.

Splitting equation (6.21) up in two finite sums, [4] and evaluating we get

$$w_{k+1} = w_0 - \eta E' \left((k+1) \frac{1-\alpha^{k+1}}{1-\alpha} - \frac{\alpha(1-\alpha^k)}{(1-\alpha)^2} + \frac{k\alpha^{k+1}}{1-\alpha} \right)$$

$$= w_0 - \eta E' \left(\frac{k+1}{1-\alpha} \right) \left(1 - \frac{1-\alpha^{k+1}}{k+1} \frac{\alpha}{1-\alpha} \right). \tag{6.22}$$

The effect of α is now clear. In flat regions of weight space the convergence rate is accelerated with a factor approaching $1/(1-\alpha)$, when k gets large. In narrow steep regions the effect of α is to average out components of the gradient with alternating signs.

As we shall see in section 6.5, gradient descent with momentum is an approximation of a conjugate gradient update. In conjugate gradient methods, however, η and α are chosen automatically. The problem with the gradient descent with momentum is that η and α has to be guessed by the user. Furthermore, the optimal values of η and α might change in each iteration.

6.4.4 Adaptive learning rate and momentum

Many heuristic schemes to adapt the learning rate and/or the momentum dynamically have been proposed in the literature, such as [131, 147, 134, 170, 97, 124, 87, 211]. It is impossible to describe them all here. We will, however, briefly describe some of the best known approaches.

One heuristic to adapt both learning rate and momentum has been proposed by Chan and Fallside [134]. The main idea of the learning rate adaptation is to calculate the angle θ_k between the current gradient and the last weight update and use this as information about the error surface characteristics. If $90° \le \theta_k \le 270°$, arrival at a ravine wall is likely and the learning rate should be decreased. If θ_k approaches $0°$ or $360°$, arrival at a plateau is likely and the learning rate should be increased. Chan and Fallside suggest the following adaptation of η

$$\eta_k = \eta_{k-1}(1 + \tfrac{1}{2}\cos\theta_k). \tag{6.23}$$

When a constant momentum is used, the weight update vector can be dominated by the momemtum term and even point uphill instead of downhill. The idea of the adaptation of the momentum is therefore to insist on having the magnitude of the momentum term smaller than the magnitude of the gradient term. In this case the gradient term will always be the dominating factor in the weight update vector. The adaptation of the momentum is given by

$$\alpha_k = \alpha_0 \eta_k \frac{|\Delta E(w_k)|}{|\Delta w_{k-1}|} \qquad 0 < \alpha_0 < 1. \tag{6.24}$$

---

[4] Here we use that: $\sum_{j=0}^k x^j = \frac{1-x^{k+1}}{1-x}$ and $\sum_{j=0}^k jx^j = \frac{x(1-x^k)}{(1-x)^2} - \frac{kx^{k+1}}{1-x}$, $|x| < 1$.

This method yields good results compared to standard gradient descent, but is, however, not without problems. The setting of η_0 and α_0 might be crucial for the success of this adaptation scheme. Chan and Fallside incorporate a backtracking scheme into the algorithm to prevent too large learning rates caused from too high an initial η_0 value. If the error is larger than the previous error, the learning rate η_k is then reduced by a half. See [133] for a comparison of this method with other adaptive methods.

Several researchers have explored the idea of having a learning rate and/or a momentum term for each unit or even for each weight in the network. The motivation for this strategy is that parameters appropriate for any one weight dimension might not be appropriate for other dimensions. Note that having different learning rates for each unit or each weight means that the weights are not modified in the direction of the negative gradient any longer. Thus, such a system is no longer experiencing gradient descent. Instead, the weights are updated based on gradient information together with estimated information about the curvature.

One scheme, that has learning rates and momentum for each unit is described by Haffner *et al* [162]. Here heuristic schemes are developed for adaptation of both learning rate and momentum. The idea for the learning rate scheme is to limit the norm of the learning rate times the gradient to a fixed value, say ω. This can be achieved by adapting the learning rate η_m^l associated with unit number m in layer l with

$$\eta_m^l = \frac{\eta}{1 + \eta/\omega \sqrt{\sum_{n_s^r \in S_m^l} \left(dE/dw_{ms}^{lr} \right)^2}} \qquad \eta > 0 \qquad \omega > 0. \qquad (6.25)$$

Haffner *et al* state that a value of ω equal to one yields good results. However, they do not say anything about the value of the other user-dependent parameter η, the value of which might be crucial for the convergence rate.

The momentum term is adapted in a similar fashion. The overall idea is, that the more the network changes the smaller should the momentum be. A characteristic symptom of too large a momentum is divergence of the term $|w|^2$ over time. Thus in the kth iteration and for each unit, Haffner *et al* define a control measure by

$$Q_m^l(k) = \sum_{n_s^r \in S_m^l} [w_{ms}^{lr}(k)]^2 - \sum_{n_s^r \in S_m^l} [w_{ms}^{lr}(k-1)]^2$$

$$= 2 \sum_{n_s^r \in S_m^l} w_{ms}^{lr}(k-1) \Delta w_{ms}^{lr}(k) + \sum_{n_s^r \in S_m^l} [\Delta w_{ms}^{lr}(k)]^2. \qquad (6.26)$$

In order to limit $Q_m^l(k)$, the momentum is chosen such as to be inverse proportional to the first term in (6.26). The adaptation formula is

$$\alpha_m^l = \frac{1}{1 + \psi | \sum_{n_s^r \in S_m^l} w_{ms}^{lr}(k-1) \Delta w_{ms}^{lr}(k) |} \qquad \psi > 0. \qquad (6.27)$$

This adaptation formula does not ensure that the control measure $Q_m^l(k)$ is limited to a fixed value as was the case for learning rate adaptation. Through several experiments Haffner *et al* conclude that the method yields faster convergence than standard gradient descent and also gives a higher percentage of converging trials.

Another heuristic method of adapting learning rates is the *delta-bar–delta* method proposed by Jacobs [170]. In this case there are independent learning rates for each single weight. Jacobs develops a gradient-descent-like updating rule for the learning rates for each unit in the network. Let Θ be a diagonal matrix of learning rate values. Then we can approximate the derivative $dE/d\Theta$ by

$$\frac{dE}{d\Theta} \approx \frac{dE}{dw_k^T}\frac{dw_k}{d\Theta}. \tag{6.28}$$

The derivative with respect to each learning rate η_{ms}^{lr} is then given by

$$\frac{\delta E}{\delta \eta_{ms}^{lr}} = -\frac{\delta E}{\delta w_{ms}^{lr}(k)}\frac{\delta E}{\delta w_{ms}^{lr}(k-1)}. \tag{6.29}$$

$dE/d\Theta$ can then be updated simultanously with the weights by the gradient descent update rule

$$\Delta \eta_{ms}^{lr} = \gamma \frac{\delta E}{\delta w_{ms}^{lr}(k)}\frac{\delta E}{\delta w_{ms}^{lr}(k-1)} \qquad \gamma > 0. \tag{6.30}$$

This update scheme is called the *delta–delta* rule and is, unfortunately, of limited practical use. The problem is, that the convergence of the process is crucially dependent on the value of γ. Jacobs overcomes this problem by defining a new update rule, which only in principle works in a similar fashion to the delta–delta rule. The delta-bar–delta rule is given by

$$\Delta \eta_{ms}^{lr} = \begin{cases} \gamma & \bar{\delta}_{k-1}\delta_k > 0 & \gamma > 0 \\ -\phi \eta_{ms}^{lr} & \bar{\delta}_{k-1}\delta_k < 0 & \phi > 0 \\ 0 & \text{otherwise} \end{cases} \tag{6.31}$$

where $\delta_k = \delta E/(\delta w_{ms}^{lr}(k))$ and $\bar{\delta}_k = (1-\theta)\delta_k + \theta\bar{\delta}_{k-1}, 0 < \theta < 1$, i.e. $\bar{\delta}_k$ is a running average of the current and past gradients. If the current gradient has opposite sign as the running average gradient the learning rate is decreased exponentially. If the current gradient has the same sign as the running average gradient then the learning is increased linearly. The difference between the delta–delta rule and the delta-bar–delta rule is that the latter takes average gradients into account and updates the learning rates independently of the size of the current gradient.

Jacobs reports a significant increase of convergence compared to standard gradient descent. There are, however, some problems with the delta-bar–delta method that are unclarified. A problem that can be immediately identified, is how to select the values of the two user-dependent parameters γ and ϕ. The values of these parameters might be very crucial for the success of this scheme. Jacobs does not give any description of how to set these parameters.

6.4.5   Learning rate schedules for on-line gradient descent

In this section we present some promising learning rate schedules introduced by Darken *et al* [140]. These schedules are only functions of time and not of previous values of learning rates or other parameters. The schedules are based on results from stochastic approximation theory, see for example [201] and [159].

In standard stochastic approximation theory results are given about the convergence properties of on-line gradient descent. On-line gradient descent on the least-mean-square-error function is guaranteed to converge if the learning rate η_k satisfies

$$\sum_{k=1}^{\infty} \eta_k = \infty \quad \text{and} \quad \sum_{k=1}^{\infty} \eta_k^2 = 0. \quad (6.32)$$

When the learning rate η_k is only a function of time, it can be shown that the optimal rate of convergence is proportional to k^{-1}, i.e. $|w_k - w_*|^2 \propto k^{-1}$, where w_* is the desired minimum. When the learning rate is allowed to depend on current or previous values of the learning rate or other parameters, as in the adaptive schemes described in the last section, very little is known theoretically about the optimal convergence rate. In [159] it is shown that in order to converge at an optimal rate, we must have $\eta_k \to c/k$ asymptotically, for c greater than some threshold c_*, which depends on the error function and on the training set. Chung and Lee show that c_* is equal to $1/2\lambda_{min}$, where λ_{min} is the smallest eigenvalue of the Hessian of the error function [135]. The usual choice of learning rate schedule in stochastic approximation theory is $\eta_k = c/k$. However, this scheme often converges slowly. Darken *et al* propose a more sophisticated schedule that guarantees an asymptotically optimal rate of convergence. The schedule is called *search-then-converge (STC)* and is given by

$$\eta_k = \eta_0 \frac{1 + (c/\eta_0)(k/\tau)}{1 + (c/\eta_0)(k/\tau) + \tau k^2/\tau^2}. \quad (6.33)$$

The main idea of this schedule is to delay the major decrease in the learning rate until a minimum has been located. The parameter η_k is approximately equal to η_0; when it is small compared to $\sqrt{\tau}$ this is called the 'search phase'. For times when it is greater than $\sqrt{\tau}$, the learning rate decreases as c/k, which is called the 'convergence phase'. Darken *et al* demonstrate major improvements in convergence rate using this schedule compared to traditional learning rate schedules. There are, however, some problems with the method of setting initial parameters such as c, η_0 and τ. Darken *et al* address the problem of setting the c parameter. As mentioned above c should be greater than $c_* = 1/2\lambda_{min}$. In fact, Darken *et al* show that the system exhibits a kind of *phase transition* at

$c = c_*$. This means that an arbitrarily small change in c, which moves it to the opposite side of c_* has a dramatic effect on the behaviour of the system. Darken *et al* argue that using a direct method of estimating c_* is too time consuming since this involves estimation of the smallest eigenvalue of the Hessian. For this reason, they outline an *ad hoc* method of determining whether a particular value of c is less than c_*. They use a heuristic scheme to adapt c based on characteristics of the weight vector trajectory. It is, however, in our opinion an open question whether the direct method of estimating the smallest eigenvalue is too time consuming after all. The smallest eigenvalue of the Hessian can be estimated by use of the Power method on the matrix $B = (\beta I - E''(\boldsymbol{w}_k))$, where $\beta > \lambda_{max}$ [200]. This estimation can be relaxed to an on-line situation by replacing B with a running average of the form

$$B_k = (1 - \gamma)B_{k-1} + \gamma(\beta I - E_p''(\boldsymbol{w}_k)) \qquad 0 < \gamma < 1. \qquad (6.34)$$

If the Power method is run simultanously with the update of weights, then this scheme would cost $O(N)$ times more per weight update, i.e. twice as much calculation work per iteration [197, 188].

Darken *et al* do not address the setting of the parameters η_0 and τ, because these values do not affect the asymptotic behaviour of the system. The values do, however, have a major affect on the rate of convergence and methods for the initial setting of these parameters are needed in order for the method to have any real practical use. Some practical hints about how to set these parameters include the following [139]. The parameter η_0 should be as large as possible (but avoiding instability). The parameter $\sqrt{\tau}$ should be some fraction of the total number of iterations that are planned to be run. In the setting of τ, prior knowledge about the problem can be used. This could be knowledge about how noisy the problem is, or if it is easy to locate a minimum and so forth.

6.4.6   The quickprop method

Quickprop is not strictly a gradient descent method, but can be viewed somewhere between a gradient descent method and a Newton method [143]. It has, in our opinion not received the attention it deserves. We give a detailed description here.

In order to avoid the time and space consuming calculations involved with the Newton algorithm two approximations are made. The Hessian matrix is approximated by ignoring all non-diagonal terms, making the assumption that all the weights are independent. Each term in the diagonal is approximated by a one-sided difference formula given by

$$\frac{\mathrm{d}^2 E(w)}{\mathrm{d}w^2} \approx \frac{E'(w_k) - E'(w_{k-1})}{w_k - w_{k-1}} \qquad (6.35)$$

where w_k is a given weight at time step k. The value $\mathrm{d}^2 E(w)/\mathrm{d}w^2$ can actually be calculated precisely with a little more calculation work [176], but is in the quickprop method not more efficient than the approximation. Geometrically the two approximations can be interpreted as follows. The error versus weight curve for each weight is approximated by a quadratic, which is assumed to be independent of changes in other weights. The main idea is to compute the minimum of this quadratic for each weight and update the weights by the following formula

$$\Delta w_k = -(\eta_k E'(w_k) + \alpha_k) \tag{6.36}$$

where η_k is

$$\eta_k = \begin{cases} \eta_0 & E'(w_k)E'(w_{k-1}) > 0 \\ 0 & \text{otherwise} \end{cases} \tag{6.37}$$

and α_k is

$$\alpha_k = \begin{cases} \frac{w_k - w_{k-1}}{E'(w_k) - E'(w_{k-1})} E'(w_k) & \left| \frac{E'(w_k) - E'(w_{k-1})}{E'(w_{k-1})} \right| < \frac{\mu}{1+\mu} \\ \mu \Delta w_{k-1} & \text{otherwise.} \end{cases} \tag{6.38}$$

The constant η_0 is similar to the learning rate in gradient descent. If $E'(w_k)E'(w_{k-1}) > 0$, i.e. the minimum of the quadratic has not been passed, a linear term is added to the quadratic weight change. On the other hand, if $E'(w_k)E'(w_{k-1}) \leq 0$, i.e. the minimum of the quadratic has been passed, only the quadratic weight change is used to go straight down to the minimum. The parameter μ is usually set equal to 2, which seems to work well in most applications.

The algorithm is usually combined with a *prime-offset* term added to the first derivative of the sigmoid activation function. The use of this term can influence the quality of the solutions found. Despite the two very crude approximations the quickprop algorithm has shown very good performance in practice. One drawback with the algorithm is, however, that the η_0 parameter is very problem dependent. An adaptive scheme to estimate this parameter would significantly increase the usefulness of this method. It might be possible to combine one of the adaptive learning rate schemes, described above, to adapt η_0.

6.4.7 Estimation of optimal learning rate and reduction of large curvature components

The eigenvalues of the Hessian matrix can be interpreted as the curvature in the direction of the corresponding eigenvectors. The eigenvectors are the main axes of the contours of equal error, which are approximately ellipsoids with the minimum of the error as a centre. Only if all the eigenvalues are of equal size, does the gradient descent direction point directly to the minimum (see figure 6.1). So, as concluded in the analysis of the convergence rate of the gradient descent method, the main factor limiting the convergence rate of gradient descent is that the curvature of the error has different values in different directions. The largest curvature limits the maximum value of the learning rate, whilst the smallest curvature dominates the learning time. The optimal learning rate for off-line gradient descent can be shown to be equal to the inverse of the largest eigenvalue. In this section we describe methods to reduce the influence of large curvature components and also how to estimate the optimal learning rate.

The true direction to the minimum can be computed by multiplying the gradient descent vector by the inverse of the Hessian matrix, assuming that the Hessian is invertible. We then get the Newton direction (see section 6.6). The inverse of the Hessian can be expressed in terms of the eigenvectors and corresponding eigenvalues. By the spectral theorem from linear algebra we have that $E''(w_k)$ has N eigenvectors that form an orthogonal basis in $\Re^N$ [169]. This implies that the inverse of the Hessian matrix $E''(w_k)^{-1}$ can be written in the form

$$E''(w_k)^{-1} = \sum_{i=1}^{N} \frac{e_i e_i^T}{|e_i|^2 \lambda_i} \tag{6.39}$$

where λ_i is the ith eigenvalue of $E''(w_k)$ and e_i is the corresponding eigenvector. Equation (6.39) implies that the Newton search directions d_k can be written as

$$d_k = -E''(w_k)^{-1} E'(w_k) = -\sum_{i=1}^{N} \frac{e_i e_i^T}{|e_i|^2 \lambda_i} E'(w_k)$$

$$= -\sum_{i=1}^{N} \frac{e_i^T E'(w_k)}{|e_i|^2 \lambda_i} e_i \tag{6.40}$$

where $E'(w_k)$ is the gradient vector. So the Newton search direction can be interpreted as a sum of projections of the gradient vector onto the eigenvectors weighted with the inverse of the eigenvalues. To calculate all eigenvalues and corresponding eigenvectors costs $O(N^3)$ time which is infeasible for large N. Le Cun *et al* [216] argue that only a few of the largest eigenvalues and the corresponding eigenvectors are needed to achieve a considerable speed up in learning. The idea is to reduce the weight change in directions with large curvature, while keeping it large in all other directions. A choice of search direction could be

$$d_k = -\left(E'(w_k) - \mu \sum_{i=1}^{k} \frac{e_i^T E'(w_k)}{|e_i|^2} e_i \right) \tag{6.41}$$

$$0 \le \mu \le 1$$

where i now runs from the largest eigenvalue λ_1 down to the kth largest eigenvalue λ_k, and μ is some appropriate constant (Le Cun *et al* [216] suggest $\mu = \lambda_{k+1}/\lambda_1$). Equation (6.41) reduces the component of the gradient along the directions with large curvature. See also [108] for a discussion of this. The learning rate can now be increased with a factor of λ_1/λ_{k+1}, since the components in directions with large curvature have been reduced with the inverse of this factor.

Another approach also proposed by Le Cun *et al* is to use a small part of the sum in equation (6.40) as search direction, so that

$$d_k = -\sum_{i=1}^{k} \frac{e_i^T E'(w_k)}{|e_i|^2 \lambda_i} e_i \tag{6.42}$$

with $k \ll N$. In theory, this can accelerate the convergence by a factor λ_1/λ_{k+1}, compared to standard gradient descent.

The largest eigenvalue and the corresponding eigenvector can be estimated by an iterative process known as the *Power method* [200]. The Power method can be used successively to estimate the k largest eigenvalues if the components in the directions of already estimated eigenvectors are substracted in the process. Below we show an algorithm for estimation of the ith eigenvalue and eigenvector. The Power method is here combined with the *Rayleigh quotient technique* [200]. This can accelerate the process considerably.

Choose an initial random vector e_i^0. Repeat the following steps for $m = 1, \ldots, M$, where M is a small constant:

$$e_i^m = E''(w_k)e_i^{m-1} \qquad e_i^m = e_i^m - \sum_{j=1}^{i-1} \frac{e_j^T e_i^m}{|ej|^2} e_j$$

$$\lambda_i^m = \frac{(e_i^{m-1})^T e_i^m}{|e_i^{m-1}|^2} \qquad e_i^m = \frac{1}{\lambda_i^m} e_i^m.$$

Here λ_i^M and e_i^M are respectively the estimated eigenvalue and eigenvector. Theoretically it would be enough to subtract the component in the direction of already estimated eigenvectors once, but in practice roundoff errors will generally introduce these components again. The term $E''(w_k)e_i^m$ can be approximated by a one-sided difference equation of the form

$$E''(w_k)e_i^m \approx \frac{E'(w_k + \sigma e_i^m) - E'(w_k)}{\sigma} \tag{6.43}$$

$$0 < \sigma \ll 1.$$

See [189] for an explanation. It has recently been shown independently by Pearlmutter and Moller that the term also can be calculated exactly in the same order of time as the approximation [197, 188].

The method of large curvature reduction can be relaxed to an on-line situation by replacing all terms of the form $E''(w)v$, where v is a vector, with a running average of the form

$$E''(w)v = (1 - \gamma)E''(w)v + \gamma E''_p(w)v \qquad 0 < \gamma < 1 \tag{6.44}$$

where $E''_p(w)$ is the Hessian matrix associated with pattern p. Le Cun *et al* report significant increase in convergence even if only a few eigenvectors are used.

Instead of changing the search direction d_k, one can of course also use the above techniques to estimate the optimal learning rate $\eta = \lambda_{max}^{-1}$ for gradient descent. Le Cun *et al* describe a technique based on equation (6.44) to estimate λ_{max}^{-1} and use this as an estimate of the optimal on-line learning rate. It should here be noted, that the learning rate $\eta = \lambda_{max}^{-1}$ is not necessarily an optimal choice in the on-line version of gradient descent. Le Cun *et al* report, however, very impressive results with this scheme. The on-line estimation of the largest eigenvalue based on (6.44), seems to converge very quickly, i.e. after presentation of a small fraction of the whole training set. The largest eigenvalue seems to be mainly determined by network architecture and initial weights, and by low-order statistics of the training data.

6.5   Conjugate gradient

Conjugate gradient methods can be regarded as being somewhat intermediate between the method of gradient descent and Newton's method described in section 6.6. They are motivated by the desire to accelerate the typically slow convergence associated with the gradient descent method while avoiding the information requirements associated with the evaluation, storage and inversion of the Hessian matrix as required by the Newton method. The standard conjugate gradient method was originally developed by Hestenes and Stiefel to solve a set of equations with a positive definite matrix of coefficients [166]. Since then, conjugate gradient methods have become a standard method for non-linear function minimization.

In this section we give a brief introduction to the conjugate gradient methods and describe the convergence properties of the methods. The scaled conjugate gradient method is described in the last part of this section.

6.5.1   Non-interfering directions of search

One of the problems with the gradient descent method was that the gradient descent directions were interfering, so that a minimization in one direction could spoil past minimizations in other directions. This problem is solved in the conjugate gradient methods and is the heart of these methods. Under the assumption that the error function is quadratic, conjugate gradient methods produce non-interfering directions of search. This implies that in the kth iteration, the error has been minimized over the whole subspace spanned by all previous search directions. The necessary and sufficient condition to have non-interfering directions of search is, that the directions have to be mutually conjugate with respect to the Hessian matrix. So if $p_1, p_2, \ldots, p_N$ is a set of directions, we have

$$p_i^T E''(w)p_j = 0 \qquad \text{when} \qquad i \neq j. \qquad (6.45)$$

This is easy to verify by the following. If we minimize the error optimally in the direction of say p_i, then we have

$$\frac{d}{d\alpha} E(w_i + \alpha p_i) = 0 \Rightarrow E'(w_{i+1})^T p_i = 0. \qquad (6.46)$$

In order to keep the error minimized in this direction, equation (6.46) has to be satisfied for all coming minimizations. So after minimization in a new direction p_j, we need the condition

$$E'(w_{j+1})^T p_i = 0 \qquad (6.47)$$

to be satisfied. Figure 6.2 illustrates the situation.

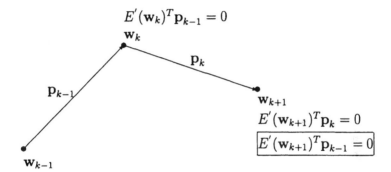

Figure 6.2. Non-interfering directions p_{k-1} and p_k. The formula in the box indicates the non-interference condition.

So we need to show that (6.45) $\Leftrightarrow$ (6.47). This can be shown by induction. Assume that equation (6.47) is satisfied for all $w_k, k < j + 1$. The initial step, $k = i + 1$, is immediately true by equation (6.46). Now $E(w_{j+1})$ can be approximated by

$$E(w_{j+1}) = E(w_j + \alpha_j p_j) \approx E(w_j) + \alpha_j p_j^T E'(w_j)$$
$$+ \tfrac{1}{2}\alpha_j^2 p_j^T E''(w_j)p_j. \qquad (6.48)$$

Observe that the α_j that minimizes the quadratic (6.48) is given by

$$\alpha_j = \frac{-E'(w_j)^T p_j}{p_j^T E''(w_j)p_j}. \qquad (6.49)$$

Differentiation of (6.48) gives

$$E'(w_{j+1}) \approx E'(w_j) + \alpha_j E''(w_j)p_j. \qquad (6.50)$$

Multiplying by p_i gives the desired result

$$E'(w_{j+1})^T p_i = 0$$
$$\Leftrightarrow \left(E'(w_j) + \alpha_j p_j^T E''(w_j) \right)^T p_i = 0$$
$$\Leftrightarrow p_j^T E''(w_j) p_i = 0 \qquad (6.51)$$

The search directions can be determined recursively such that (6.45) is satisfied. The idea is to choose p_k to be the projection of the current gradient descent vector onto the subspace orthogonal to the subspace spanned by the previous directions $p_1, p_2, \ldots, p_{k-1}$. The following lemma describes how [146].

Lemma 6.4. Assume that the error function is quadratic with constant Hessian $E''(w)$. Let the direction vectors be recursively defined by

$$p_k = -E'(w_k) + \beta_{k-1} p_{k-1} \qquad (6.52)$$

where

$$\beta_{k-1} = \frac{|E'(w_k)|^2 - E'(w_k)^T E'(w_{k-1})}{|E'(w_{k-1})|^2} \qquad \beta_{-1} = 0. \qquad (6.53)$$

Then the following three conditions hold

$$p_k^T E''(w) p_i = 0 \qquad i < k \qquad (6.54)$$
$$E'(w_k)^T E'(w_i) = 0 \qquad i < k \qquad (6.55)$$
$$-E'(w_k)^T p_k = |E'(w_k)|^2. \qquad (6.56)$$

The conditions are referred to as mutual conjugacy, orthogonal gradient and descent conditions.

Proof. We prove the lemma by induction. The initial step for p_0 and p_1 is easy to verify using the fact that $p_0 = -E'(w_0)$. We leave that to the reader. Assume that the lemma is true for all $i < k$. We first prove the orthogonal gradient condition. Using (6.50) and (6.52) we have

$$E'(w_k)^T E'(w_i) = \left(E'(w_{k-1}) + \alpha_{k-1} p_{k-1}^T E''(w) \right)^T E'(w_i)$$
$$= E'(w_{k-1})^T E'(w_i) + \alpha_{k-1} p_{k-1}^T E''(w) \left(\beta_{i-1} p_{i-1} + p_i \right).$$

When $i < k - 1$, this is zero by (6.54) and (6.55), and when $i = k - 1$, it is zero by (6.49), (6.54) and (6.56). Thus, the orthogonal gradient condition (6.55) is true. Using (6.52) and (6.50)

$$p_k^T E''(w) p_i = (-E'(w_k) + \beta_{k-1} p_{k-1})^T E''(w) p_i$$
$$= -\frac{1}{\alpha_i} E'(w_k) \left(E'(w_{i+1}) - E'(w_i) \right) + \beta_{k-1} p_{k-1}^T E''(w) p_i.$$

$$(6.57)$$

When $i < k - 1$, this is zero by (6.54) and (6.55), and when $i = k - 1$, it is zero by (6.49), (6.53), (6.55) and (6.56). So the mutual conjugacy condition is true. Finally, by (6.52) and (6.46)

$$-E'(w_k)^T p_k = -E'(w_k)^T \left(-E'(w_k)^T + \beta_{k-1} p_{k-1}\right)$$
$$= E'(w_k)^T E'(w_k). \tag{6.58}$$

Thus, the descent condition is true, which ends the proof.

Lemma 6.4 does not hold for non-quadratic functions. However, the direction vectors produced by lemma 6.4 will be approximately non-interfering, since a non-quadratic error function can be approximated by a quadratic as in (6.48). Based on lemma 6.4 the standard conjugate gradient method can be formulated as follows.

1. Select initial weight vector w_0;
   $r_0 = -E'(w_0)$;
   $p_0 = r_0$;
2. $\alpha_k = \min_\alpha E(w_k + \alpha p_k)$;
3. $w_{k+1} = w_k + \alpha_k p_k$;
   $r_{k+1} = -E'(w_{k+1})$;
4. if (k mod $N = 0$) then
      $p_{k+1} = r_{k+1}$;
   else
      $\beta_k = |r_{k+1}|^2 - r_{k+1}^T r_k / |r_k|^2$;
      $p_{k+1} = r_{k+1} + \beta_k p_k$;
5. $k = k + 1$; terminate or go to 2.

The learning rate in step 2 is usually determined by a one-dimensional line search, which can be very time consuming. The scaled conjugate gradient algorithm described at the end of this section, avoids this line search and estimates the learning rate by use of formula (6.49) and a scaling mechanism. If the conjugate gradient method is performed on a quadratic function, the method will terminate in at most N iterations, since by then, the whole weight space has been minimized, because of the non-interference condition. When the method is used on non-quadratic functions, this might not be the case, and the search direction is reset to the gradient descent direction. In practice, however, the method often terminates in $i \ll N$.

6.5.2 Convergence rate

We now turn to the convergence rate of conjugate gradient methods. In order to be able to say anything about the convergence rate, we again have to assume that the error function is quadratic and additionally that the Hessian matrix is constant over time. We refer to the Hessian by $E''(w)$. Thus, the bounds that we present are not strictly valid for non-quadratic error functions, but merely approximations.

The convergence rate depends very much on the distribution of the eigenvalues of the Hessian matrix. If the eigenvalues fall into multiple or close groups, the convergence rate will be fast. This can be realized by the following. Using (6.52) and the relation

$$E'(w_{k+1}) \approx E'(w_k) + \alpha_k E''(w) p_k \tag{6.59}$$

we see, that the conjugate direction vectors

$$p_k \in K_k(E''(w), E'(w_0)) \tag{6.60}$$

where

$$K_k(E''(w), E'(w_0)) = \text{span}(E'(w_0), E''(w)E'(w_0), \dots, \tag{6.61}$$
$$E''(w)^{k-1} E'(w_0))$$

is the *Krylov subspace* [98]. The weight vectors, generated in the conjugate gradient algorithm, then have the folllowing property.

$$w_k = w_0 + P_k(E''(w))E'(w_0)$$
$$\in w_0 + K_k(E''(w), E'(w_0)) \tag{6.62}$$

where $P_k(E''(w))$ is a matrix polynomial of degree k. Corresponding to $E'(w_0)$ there are uniquely determined eigenvalues $\lambda_1 < \lambda_2 < \cdots < \lambda_m$ and normalized eigenvectors $e_1, \cdots, e_m$ of $E''(w)$ such that

$$E'(w_0) = \sum_{i=1}^{m} \mu_i e_i. \tag{6.63}$$

These eigenvalues and eigenvectors are the *active* ones. The possible other eigenvalues and eigenvectors do not participate in the conjugate gradient process. Obviously the maximum dimension of the Krylov subspace for increasing k is equal to m, and the conjugate gradient algorithm will terminate in m iterations. If the Hessian only has a few distinct eigenvalues, then m can be expected to be small.

As was the case for the gradient descent algorithm, the convergence rate can be bounded by the condition number of the Hessian. The following lemma states how.

Lemma 6.5. Assume that the error function is quadratic with constant Hessian $E''(w)$. At every step in the conjugate gradient algorithm it holds that

$$\frac{E(w_{k+1}) - E(w_*)}{E(w_k) - E(w_*)} \leq 4 \left(\frac{\sqrt{\kappa} - 1}{\sqrt{\kappa} + 1} \right)^2$$

where κ is the condition number of $E''(w)$, and w_* is the minimum of the error.

Lemma 6.5 predicts only linear convergence by a fixed factor $((\sqrt{\kappa} - 1)/(\sqrt{\kappa} + 1))^2$. In order to state something about the so-called superlinear convergence behaviour, it is necessary to look further into the distribution of the eigenvalues. We refer to [98] for a more detailed evaluation of the convergence behaviour of conjugate gradient methods.

6.5.3 Scaled conjugate gradient

This section describes the scaled conjugate gradient algorithm (SCG) developed by Moller [189]. The main ideas of the algorithm are described together with a pseudo-code version of the algorithm. Then a discussion of various possible improvements of the algorithm is given.

The estimation of the learning rate in the standard conjugate gradient algorithm is done with a line search routine. A line search performs a one-dimensional iterative search for a learning rate in the direction of the current search direction. There are several drawbacks to doing a line search. It introduces new problem-dependent parameters, e.g. a parameter to determine how many iterations to perform before termination. Kinsella has shown that this can have a major impact on the performance of the conjugate gradient algorithm [171]. Furthermore, the line search does in each iteration involve several calculations of the error and/or the derivative to the error, which is time consuming.[5] SCG substitutes the line search by a scaling of the step that depends on success in error reduction and goodness of the quadratic approximation to the error. The algorithm encorporates ideas from the *model-trust region* methods in optimization and 'safety' procedures that are absent in standard conjugate gradient.

If the error function were strictly quadratic with a positive definite Hessian matrix, the learning rate given by formula (6.49) would be optimal. Using this formula for non-quadratic error functions, however, causes problems, because the Hessian matrix can be indefinite and the quadratic approximation to the error might not always be good. The key idea of SCG consists of the introduction of a scalar, λ_k, which is used to regulate the positive definiteness of the Hessian.[6] The Hessian is substituted with

$$E''(w_k) + \lambda_k I \tag{6.64}$$

so that the quadratic approximation of the error on which the minimization is performed is

$$\begin{aligned} E(w_k + \alpha_k p_k) &\approx E(w_k) + \alpha_k p_k^T E'(w_k) \\ &+ \alpha_k^2 p_k^T (E''(w_k) + \lambda_k I) p_k. \end{aligned} \tag{6.65}$$

---

[5] Each calculation costs $O(PN)$ time.
[6] This is essentially a Levenberg–Marquardt approach [146].

The learning rate that minimizes (6.65) is now given by

$$\alpha_k = \frac{-E'(w_k)p_k}{p_k^T E''(w_k)p_k + \lambda_k |p_k|^2}. \tag{6.66}$$

The values of λ_k directly scale the step size in such a way that the bigger λ_k is the smaller the step size, which agrees well with our intuition of the function of λ_k.

The positive definiteness of the augmented Hessian in (6.64) is controlled in the following manner. Let the scalar δ_k be defined as

$$\delta_k = p_k^T s_k = p_k^T (E''(w_k) + \lambda_k I)p_k \tag{6.67}$$

where $s_k = (E''(w_k) + \lambda_k I)p_k$. δ_k directly reveals if $E''(w_k)$ is not positive definite. If $\delta_k \leq 0$ then the Hessian is not positive definite and λ_k is raised and s_k is estimated again. If the new s_k is renamed as $\bar{s}_k$ and the raised λ_k as $\bar{\lambda}_k$ then $\bar{s}_k$ is

$$\bar{s}_k = s_k + (\bar{\lambda}_k - \lambda_k)p_k. \tag{6.68}$$

Assume in a given iteration that $\delta_k \leq 0$. It is possible to determine how much λ_k should be raised in order to get $\delta_k > 0$. If the new δ_k is renamed as $\bar{\delta}_k$ then

$$\begin{aligned}
\bar{\delta}_k &= p_k^T \bar{s}_k = p_k^T (s_k + (\bar{\lambda}_k - \lambda_k)p_k) \\
&= \delta_k + (\bar{\lambda}_k - \lambda_k)|p_k|^2 > 0 \Rightarrow \bar{\lambda}_k > \lambda_k - \frac{\delta_k}{|p_k|^2}.
\end{aligned} \tag{6.69}$$

(6.69) implies that if λ_k is raised by more than $-\delta_k/|p_k|^2$ then $\bar{\delta}_k > 0$. The question is: how much should $\bar{\lambda}_k$ be raised to get an optimal solution? This question can not yet be answered, but it is clear that $\bar{\lambda}_k$ in some way should depend on λ_k, δ_k and $|p_k|^2$. A choice found to be reasonable is

$$\bar{\lambda}_k = 2\left(\lambda_k - \frac{\delta_k}{|p_k|^2}\right). \tag{6.70}$$

This leads to

$$\begin{aligned}
\bar{\delta}_k &= \delta_k + (\bar{\lambda}_k - \lambda_k)|p_k|^2 = \delta_k + \left(2\lambda_k - 2\frac{\delta_k}{|p_k|^2} - \lambda_k\right)|p_k|^2 \\
&= -\delta_k + \lambda_k |p_k|^2 > 0.
\end{aligned} \tag{6.71}$$

The quadratic approximation in (6.65) may not always be a good approximation to $E(w)$ since λ_k scales the Hessian matrix in an artificial way. A mechanism to raise and lower λ_k is needed which gives a good approximation, even when the Hessian is positive definite. An increase or decrease of the scaling parameter λ_k is controlled by the value of the variable Δ_k given by

$$\begin{aligned}
\Delta_k &= \frac{\text{real error change}}{\text{predicted error change}} \\
&= \frac{\left(p_k^T E''(w_k)p_k + \lambda_k |p_k|^2\right)(E(w_k) - E(w_{k+1}))}{\left(-E'(w_k)^T p_k\right)^2}.
\end{aligned} \tag{6.72}$$

Δ_k measures the ratio between the real error change and the predicted quadratic error change. The term $E''(w_k)p_k$ in (6.66) can either be approximated by a one-sided difference equation of the form

$$E''(w_k)p_k \approx \frac{E'(w_k + \sigma_k p_k) - E'(w_k)}{\sigma_k} \tag{6.73}$$

$$\sigma_k = \frac{\epsilon}{|p_k|^2} \qquad 0 < \epsilon \ll 1$$

or calculated exactly as described in [188] or [197]. The exact calculation is only a bit more complicated than the approximation and both schemes cost $O(PN)$ time, where P is the number of patterns in the training set and N is the number of weights. Parameter λ_k is now raised and lowered following the formula

- if $\Delta_k > 0.75$ then $\lambda_k = \frac{1}{4}\lambda_k$
- if $\Delta_k < 0.25$ then $\lambda_k = \lambda_k + (\delta_k(1 - \Delta_k))/|p_k|^2$.

The formula for $\Delta_k < 0.25$ increases λ_k such that the new step size is equal to the minimum to a quadratic polynomial fitted to $E'(w_k)^T p_k$, $E(w_k)$ and $E(w_k + \alpha_k p_k)$ [107]. The SCG algorithm is as shown below.

1. Initialize algorithm:

   {

   - Choose weight vector w_1 and scalars $0 < \sigma \le 10^{-4}$, $0 < \lambda_1 \le 10^{-6}$, $\overline{\lambda}_1 = 0$.
   - Set $p_1 = r_1 = -E'(w_1)$, $k = 1$ and success = true.

   }

2. If success then calculate second-order information:

   {

   - $\sigma_k = \sigma/|p_k|$
   - $s_k = [E'(w_k + \sigma_k p_k) - E'(w_k)]/\sigma_k$
   - $\delta_k = p_k^T s_k$.

   }

3. Scale δ_k:

   {

   - $\delta_k = \delta_k + (\lambda_k - \overline{\lambda}_k)|p_k|^2$.

   }

4. If $\delta_k \le 0$ then make the Hessian matrix positive definite:

   {

   - $\overline{\lambda}_k = 2(\lambda_k - (\delta_k/|p_k|^2))$
   - $\delta_k = -\delta_k + \lambda_k|p_k|^2$
   - $\lambda_k = \overline{\lambda}_k$.

   }

5. Calculate step size:

    {

    - $\mu_k = p_k^T r_k$
    - $\alpha_k = \mu_k / \delta_k$.

    }

6. Calculate the comparison parameter:

    {

    - $\Delta_k = 2\delta_k \left(E(w_k) - E(w_k + \alpha_k p_k)\right)/\mu_k^2$.

    }

7. If $\Delta_k \geq 0$ then a successful reduction in error can be made:

    {

    - $w_{k+1} = w_k + \alpha_k p_k$
    - $r_{k+1} = -E'(w_{k+1})$
    - $\overline{\lambda}_k = 0$, success = true.
    - If $k \bmod N = 0$ then restart algorithm: $p_{k+1} = r_{k+1}$
      else create new conjugate direction:

        {

        * $\beta_k = (|r_{k+1}|^2 - r_{k+1}^T r_k)/|r_k|^2$
        * $p_{k+1} = r_{k+1} + \beta_k p_k$.

        }

    - If $\Delta_k \geq 0.75$ then reduce the scale parameter: $\lambda_k = \frac{1}{4}\lambda_k$.

    } else a reduction in error is not possible:

    {

    - $\overline{\lambda}_k = \lambda_k$
    - success = false.

    }

8. If $\Delta_k < 0.25$ then increase the scale parameter:

    {

    - $\lambda_k = \lambda_k + (\delta_k(1 - \Delta_k))/|p_k|^2$.

    }

9. If the steepest descent direction $r_k \neq 0$ then

    {

    - set $k = k + 1$
    - go to 2

    } else

    {

    - return w_{k+1} as the desired minimum.

    }

The value of σ should be as small as possible taking the machine precision into account. When σ is kept small ($\leq 10^{-4}$), experiments indicate that the value of σ is not critical for the performance of SCG. Because of that, SCG seems not to include any user dependent parameters whose values are crucial for the success of the algorithm. This is a major advantage compared to the line search based algorithms which include these kinds of parameters.

For each iteration there is one call of $E(w)$ and two calls of $E'(w)$, which gives a calculation complexity per iteration of $O(5PN)$. When the algorithm is implemented this complexity can be reduced to $O(4PN)$ because the calculation of $E(w)$ can be built into one of the calculations of $E'(w)$. In comparison with BP, SCG involves twice as much calculation work per iteration since BP has a calculation complexity of $O(2PN)$ per iteration. The calculation complexity of CGL and BFGS is about $O(3\text{--}15PN)$ since the line search, on average, involves 3–15 calls of $E(w)$ or $E'(w)$ per iteration [161].[7]

We will now turn to a discussion of possible improvements of the algorithm. The formula (6.66) for the learning rate can be viewed as a *one-step Newton line search estimation* [185]. A j-step Newton estimation is defined as

$$\alpha_j = \alpha_{j-1} + \frac{-E'(w_k + \alpha_{j-1}p_k)p_k}{p_k^T E''(w_k + \alpha_{j-1}p_k)p_k + \lambda_k|p_k|^2} \tag{6.74}$$

$$\alpha_0 = 0.$$

Using (6.74) in SCG with j as a user dependent parameter, would be to introduce a procedure similar to the line search that SCG was designed to avoid. However, it might be better to set j to a constant other than $j = 1$. To check if this is the case, a series of expriments were run on the two-spirals problem [175] with various values of j. A series of ten runs was tested for each value of j, and the runs were terminated when all the patterns were classified correctly within a margin of 0.8.[8] The results are illustrated in table 6.1.

We observe that the convergence with respect to the number of epochs improves with increasing j, except for $j = 3$. The convergence does, however, not improve enough to justify the additional computation time used in each iteration.

Peter Williams has suggested several interesting improvements to the SCG algorithm [107]. One is an improvement of the update formula of the scaling parameter λ_k, which is included in [189]. Another is an adaptive scheme for the setting of the parameter ϵ in the one-sided difference formula (6.73). We will present his findings here. In [189] it was experimentally shown that the value of ϵ was not problem dependent as long as it was set to a small value. Williams argues, however, that a small gain in convergence can be achieved by adapting ϵ. If the error were strictly quadratic, the one-sided differencing would be an exact calculation of $E''(w_k)p_k$ no matter what the value of ϵ. So

---

[7] Notice when λ_k is zero, SCG is equal to the standard conjugate gradient algorithm (CG).

[8] The *exponential* error function described in [187] was used in these experiments.

Table 6.1. Average of ten runs on the two-spirals problem with various values of the j parameter. The $\langle cu \rangle$ part is the average number of *complexity units*, which is an abstract measure that also takes the computational costs per epoch into account (one $\langle cu \rangle$ is equivalent to one forward or one backward propagation of all patterns in the training set).

j	$\langle$epoch$\rangle$	$\langle$cu$\rangle$	Failures
1	1052	4208	1
2	804	4824	0
3	840	6720	1
4	732	7320	0

in this case there is no particular advantage in having $\epsilon \ll 1$. When the error is non-quadratic, Williams argues that the value of ϵ does not matter either. When the scaling parameter $\lambda_k \approx 0$ the learning rate calculation is equivalent to fitting a straight line to $E'(w_k)^T p_k$ and $E'(w_k + \sigma_k p_k)^T p_k$ to give α_k as the estimated zero crossing of the line (see figure 6.3). The ideal σ_k would be one for which $\sigma_k = \alpha_k$. Note that α_k is the new learning rate that we want to estimate. This suggests a way of adapting ϵ. One adaptation rule given by Williams is

$$\epsilon_{k+1} = \sigma_k \left(\frac{\alpha_k}{\sigma_k} \right)^{\pi} \qquad 0 \le \pi \le 1. \qquad (6.75)$$

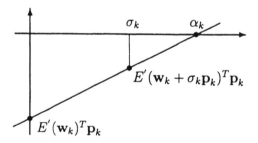

Figure 6.3. The extrapolation used to determine α_k by the one-sided differencing scheme.

Choosing $\pi = 0$ is equivalent to using the initial value ϵ_0 throughout. Non-zero values of π adapt ϵ_k so that, on average, the values of α_k and σ_k tend to be equalized. Note that if ϵ is approximately equal to the expected learning rate, then a finite difference estimate may provide a more faithful picture of the error surface than an extrapolated local quadratic model, such as the exact calculation of $E''(w_k)p_k$, especially if the quadratic approximation of the error is bad. In any case, Williams reports a minor speedup in convergence using this adaptation scheme. In [187], however, the author finds that the exact calculation of $E''(w_k)p_k$ on average yields the fastest convergence.

6.5.4 Stochastic conjugate gradient

In section 6.4 we saw that the gradient descent algorithm could be used in two different modes, *on-line* and *off-line*. Conjugate gradient algorithms are in their standard form only applicable in off-line mode, because of the near-optimal choice of learning rate in each iteration. In off-line mode the amount of computation time used in each iteration is dependent on the number of patterns in the training set. When the training set is large and contains a lot of redundant information this leads to redundant computations in an off-line algorithm. So although the convergence rate of conjugate gradient algorithms with respect to the number of iterations used is much better than that of gradient descent algorithms, it can often be observed that on-line gradient descent beats more sophisticated second-order algorithms on large redundant problems. The second-order algorithms 'drown' in their own computations. It is, however, possible to make stochastic versions of second-order algorithms that update weights based on smaller subsets of the training set. In [190] a stochastic version of the scaled conjugate gradient algorithm is described. We summarize this algorithm here. In section 6.7 on-line and off-line techniques are discussed in more detail.

The approach of updating the weights based on smaller subsets of data has been explored by several researchers. Kramer and Sangiovanni-Vincentelli describe a two-stage scheme, where the first stage involves training the network on the current subset of data until convergence and the second stage involves picking new patterns to successively increase the training subset [173]. Haffner *et al* describe a similar scheme [162]. A problem with such schemes is that the network tends to get overspecialized when trained to convergence on small subsets of data. The training should be terminated and the subset increased when overspecialization occurs. This point is, however, not easy to detect. Kuhn and Herzberg describe a scheme combined with conjugate gradient applied to a speech recognition problem, where the subset size is proportional to the number of output classes [174]. In these schemes the size of the data subsets and which patterns to include in them, is determined in a very *ad hoc* fashion, e.g. taking one pattern from each output class or adding a predetermined and constant number of patterns to the current subset. Furthermore, the schemes do not consider any validation of the size and patterns chosen, which means that the training process might diverge without detection.

Using standard sampling techniques as described in [137], it is possible to define schemes that can validate each update and base the size of the data subset on these validations. We call such schemes *update-validation*. One such scheme was described in [190], where it was combined with the scaled conjugate gradient algorithm. The scheme is illustrated in figure 6.4. The scheme involves two blocks (subsets) of data, an *update* block and a *sample* block. An update of the weights is based on the update block and the sample block is used to validate each update. This validation involves the calculation of an *update probability* $\hat{P}_i$, that is an estimate of the probability that an update will decrease the error on the whole training set. This probability can be estimated under the assumption that the error of blocks of data is normally distributed around the error of the whole training set, and then selecting the sample block as a simple random sample. The update probability can then be estimated by

$$\hat{P}_i = \frac{1}{\sqrt{2\pi}} \int_{-\infty}^{\Delta\hat{\mu}_i/\sigma_i} e^{-\frac{1}{2}t^2} dt \qquad (6.76)$$

where $\hat{\mu}_i$ is the change in error of the current sample block before and after a possible update, and σ_i^2 is the error variance of the sample block. The size of the update block is optimized by means of a binary search method. The size of the sample block is chosen such that the error of a sample block is very close to the error of the training set with a high probability.

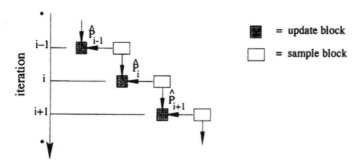

Figure 6.4. The update-validation scheme used in SSCG. $\hat{P}_i$ is an estimated probability that an update based on the current subset of data will decrease the error on the whole training set. The weights are updated with probability $\hat{P}_i$.

Only a little change in conjugate gradient algorithms is needed in order to be able to update on subsets of data of perhaps varying size. The error function has to be normalized so that we operate on average error rather than total error. If the error is normalized then the error on subsets of data will be approximations to the error on the whole training set. The better the approximations are the better and more reliable will a conjugate gradient algorithm converge. Combining the above update-validation scheme with the scaled conjugate gradient algorithm yields good results as concluded in [190]. The major advantage of combining second-order training algorithms with stochastic schemes is that a minimum is quickly localized because of more frequent updates, and the convergence is fast down to this minimum because of the second-order properties of the algorithms.

As discussed later in section 6.7, it might be possible to improve the above update-validation scheme in various ways. One idea worth mentioning is the possible use of *active data selection* techniques to determine appropriate update blocks. Rather than selecting patterns by random sampling, the training could be made more efficient by selecting patterns that maximize the information content of the update block.

6.6 Newton related methods

In this section we briefly describe the Newton method and some of its variations that are relevant in a neural network context. We conclude with a decription of the *one-step memoryless BFGS* method, which operates in O(N) time and is very similar to the conjugate gradient methods [123]. The Newton method is in its original form not applicable for training of neural networks, because it involves an inversion of the Hessian matrix, which is computational expensive. The Newton search direction p_k is defined by the linear system

$$E''(w_k)p_k = -E'(w_k) \tag{6.77}$$

which originates when minimizing the quadratic approximation to the error function. If the error is strictly quadratic and the Hessian is positive definite this algorithm converges in one iteration, starting from any initial weight vector. For non-quadratic error functions, the algorithm converges quadratically if the Hessian is positive definite and sufficiently close to the desired minimum. Whenever the Hessian is indefinite or ill conditioned severe problems arise, such as too large learning rates and numerical problems. Many modified Newton algorithms exist that incorporate techniques to ensure a sufficiently positive definite and non-singular Hessian matrix. A class of such algorithms is the *quasi-Newton* methods. Quasi-Newton methods are based on the idea of accumulating curvature information as the iterations proceed, using the observed behaviour of $E(w_k)$ and $E'(w_k)$. The methods build up an approximation of the Hessian or the inverse of the Hessian. In the following we describe how to approximate the inverse of the Hessian. In the beginning of the kth iteration of a quasi-Newton method, an approximate inverse Hessian matrix H_k is available, which is intended to reflect the inverse curvature information already accumulated. The search direction p_k is then computed from the Newton formula

$$p_k = -H_k E'(w_k). \tag{6.78}$$

The initial matrix H_0 is usually taken to be the identity matrix, so that the first iteration is equivalent with a gradient descent update. H_k is then updated recursively in each iteration so that it approximates the inverse curvature along the current search direction p_k. If we assume that the error function is quadratic, then we have by (6.50) that

$$E''(w_k)^{-1}\left(E'(w_{k+1}) - E'(w_k)\right) = \alpha_k p_k^T = (w_{k+1} - w_k).w \tag{6.79}$$

Based on (6.79), H_{k+1} is required to satisfy the so-called *quasi-Newton condition*

$$H_{k+1}\left(E'(w_{k+1}) - E'(w_k)\right) = (w_{k+1} - w_k). \qquad (6.80)$$

There are several ways to update H_k in order to satisfy this condition. We will not describe these in detail here, but only mention the one that leads us to the most effective formula, which is the Broyden–Fletcher–Goldfarb–Shanno (BFGS) formula given by

$$H_{k+1} = H_k + \left(1 + \frac{y_k^T H_k y_k}{y_k^T s_k}\right)\frac{s_k s_k^T}{s_k^T y_k} - \frac{s_k y_k^T H_k + H_k y_k s_k^T}{y_k^T s_k} \qquad (6.81)$$

where $y_k = (E'(w_{k+1}) - E'(w_k))$ and $s_k = (w_{k+1} - w_k)$. Formula (6.81) ensures that H_k is symmetric and positive definite. The BFGS algorithm involves storage of an $N \times N$ matrix and $O(PN^2)$ in computation requirements. So in its current form, the BFGS algorithm is not applicable to training of neural networks, at least not large networks. It is, however, possible to simplify formula (6.81) so that the time and memory requirements are $O(PN)$ and $O(N)$ respectively. The idea is to apply the BFGS formula to the identity matrix I, rather than to H_k. Thus H_{k+1} is determined without reference to the previous H_k, and hence the update procedure is *memoryless*. By (6.78) and setting $H_k = I$ in (6.81), the search direction is given by

$$\begin{aligned}
p_{k+1} = &- E'(w_{k+1}) \\
&- \left(1 + \frac{y_k^T y_k}{y_k^T s_k}\right)\frac{s_k^T E'(w_{k+1})s_k}{s_k^T y_k} \\
&+ \frac{y_k^T E'(w_{k+1})s_k + s_k^T E'(w_{k+1})y_k}{y_k^T s_k}.
\end{aligned} \qquad (6.82)$$

The one-step memoryless BFGS method is usually combined with a line search to estimate appropriate learning rates. See [124] for further reading about this. When exact line searches are made, i.e. an optimal learning rate is found in each iteration, the algorithm is the same as the conjugate gradient algorithm. This is easily verified by using the observation that $p_k^T E'(w_{k+1}) = 0$, which implies that $p_k^T y_k = -p_k^T E'(w_k)$. The algorithm is considered to have similar convergence properties to the conjugate gradient algorithms. This is also consistent with results reported by Battiti, who concludes that SCG and the one-step memoryless BFGS yield comparable results [123].

6.7   On-line versus off-line discussion

As mentioned in section 6.4, training methods of feed-forward neural networks can roughly be divided up into two categories, *on-line* and *off-line* techniques. There is some confusion about these terms. The term 'on-line' refers historically to methods where the weights are updated based only on information from one single pattern, while 'off-line' refers to methods where the weights are updated based on information from the whole training set. We will use the 'on-line' term in a broader sense, referencing it to methods that update weights *independent* of the training set size. Some researchers also use the terms *stochastic* and *batch* as alternative names for on-line and off-line. Several examples of both on-line and off-line methods have been described in the past sections. The gradient descent method can be used in both on-line and off-line mode. The conjugate gradient algorithms and other second-order algorithms are in standard form all off-line algorithms, but can also with some modifications be used in on-line mode as described in section 6.5.4. Off-line schemes described in the past sections include all the adaptive learning rate schemes in section 6.4.4, the quickprop method in section 6.4.6 and scaled conjugate gradient in 6.5.3. Methods that also apply to on-line mode are the learning rate schedules in section 6.4.5, the optimal learning rate estimation and reduction of large curvature components in section 6.4.7, and the stochastic conjugate gradient scheme in 6.5.4.

There are drawbacks and benefits with both types of update scheme. Off-line algorithms are easier to analyse with regard to their convergence properties, they can choose an optimal learning rate in each iteration and can yield very high accuracy solutions. They suffer, however, from the fact that the time to prepare a weight update increases with increasing training set size. This turns out to be crucial in many large-scale problems. On-line methods can be used when the patterns are not available before training starts, and a continuous adaptation to a stream of input–output relations is desired. The randomness in the updates can help escape local minima and the time to prepare a weight update is not affected by increasing training set size. On-line methods are not good enough to produce high-accuracy solutions, e.g., function approximation, and the setting of the learning rate is not well understood.

Off-line methods like standard conjugate gradient or quasi-Newton methods should in theory yield the fastest convergence. This is, however, not the case on large-scale problems that are characterized by large and very redundant training sets. The problem that arises is illustrated in figure 6.5. If many of the patterns in the training set possess redundant information the contributions to the search direction will be similar, and waiting for all contributions before updating can be a waste of time. There is obviously a tradeoff between the accuracy of the search direction and the computational costs to calculate it. In other words, redundancy in training sets produces redundant computations in off-line algorithms.

In addition to the description of the stochastic scaled conjugate gradient algorithm, [190] presents a method of measuring the amount of redundancy in training sets. We will summarize these redundancy aspects here. First it is important to recognize that the redundant computations in an off-line algorithm cannot be expected to be a constant even though the redundant information in the training set is constant. The redundant computations will also depend on the network dynamics, i.e. of the current weights. So a measure of redundancy

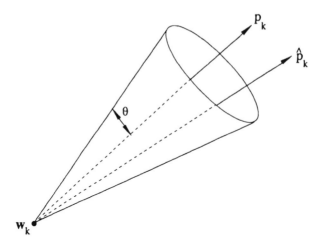

Figure 6.5. The search direction of off-line algorithms is an accumulation of partial search directions. If redundancy is present, a sum of a small fraction of these partial directions might produce a direction $\hat{p}_k$, that is 'close' to the desired search direction p_k, i.e. within a small neighbourhood of the desired direction.

of training sets alone cannot be expected to give sufficient information about the number of redundant computations made by an off-line algorithm. Such a measure could, however, give a first estimate of what to expect of the training process. A measure of redundancy can be based on standard information theory [206]. We define a redundancy measure for classification problems with M different output classes and P discrete input vectors of length L, each attribute having V possible values. The *conditional population entropy* (CPE) is defined as

$$\text{CPE} = -\sum_{m=1}^{M} p(c_m) \sum_{l=1}^{L} \sum_{v=1}^{V} p(x_v^l|c_m) \log p(x_v^l|c_m) \qquad (6.83)$$

where $p(c_m)$ is the probability that an input vector belongs to the mth class and $p(x_v^l|c_m)$ is the probability that the lth attribute of an input vector x has value v given that x belongs to the mth class. If we imagine that the whole training set is split up into M disjoint sets, one for each class, then CPE is the average information content of these sets. We can also interpret CPE as the average number of bits needed to code one input pattern given knowledge about the classes. Clearly the value of CPE states something about the degree of similarity between the input patterns. If CPE is small, only a small number of bits is needed to code the input patterns, and hence there must be great similarity between the patterns. Based on these observations, a redundancy measure RE can be defined as

$$\text{RE} = \frac{\log V - (\text{CPE}/L)}{\log V} \qquad (6.84)$$

where log V is the necessary number of bits needed to code one attribute if all values are equally likely and CPE/L is the average number of bits needed to code one attribute. One drawback with this redundancy measure is that it does not take correlations between attributes into account. Nevertheless, in [190] it is found that there is a strong correlation between the value of RE and the efficiency of various off-line and on-line algorithms, when trained on particular problems. The measure can be expanded to work on non-classification problems by splitting the input and/or the output ranges up into a set of discrete intervals.

The neural network community seems to be split up into two blocks regarding the question of which direction research should go concerning on-line and off-line algorithms. Since many practical neural network problems are characterized by large redundant training sets, some researchers think that the effort should be put into finding better speed up techniques to apply with on-line gradient descent. One such approach is the on-line estimation of learning rate and reduction of large curvature components by Le Cun *et al* described in section 6.4.7. This is a very promising approach and should be investigated further. The other group, to which the author belongs, believes that the power of second-order methods can and should be used also on large-scale problems. The approach is to make the second-order methods stochastic, i.e. perform updates on smaller blocks of data. The general approach can be characterized as follows.

- Accumulate error, gradient and/or Hessian information for a length of time that is adaptively chosen. The time interval should be large enough to ensure safe weight updates with near-optimal learning rates but small enough to avoid redundant computations.
- Use a second-order algorithm like SCG to update the weights.

One such approach is the stochastic SCG algorithm (SSCG) described in section 6.5.4. In SSCG a *validation scheme* is defined to validate each update. This scheme involves random sampling of additional patterns that are not used in the current update process. This should *in principle* not be necessary, since enough information should already be available in the data used to prepare an update. If the data are redundant, the estimated search directions produced by accumulation of partial directions will converge.[9] This means that the angle θ_k between the estimated search direction and the real direction (see figure 6.5) converges to zero. It might be possible to define an adaptive scheme that involves the convergence of θ_k to determine when to stop collecting new patterns. A problem here is that the convergence of θ_k will not necessarily be monotonic, so the scheme has to take small fluctuations of θ_k into account. These ideas are the subject for further research.

Another approach to solve the problem with redundant data is to control what data is used in training, this is often referred to as *active data selection*. This approach could be used to improve the SSCG method and other second-order stochastic methods. Rather than selecting data by random sampling, the

---

[9] We assume here that we have an infinite stream of data available.

training could be made more efficient by actively selecting data that maximize
the information density of the training subset. Active data selection has been
extensively studied in economic theory and statistics, see for example [145]. In
a neural network context Plutowski *et al* and MacKay have recently proposed
different schemes to select data [199, 181].

Plutowski *et al* consider the problem of selecting training subsets from a
large noise-free data set [199]. They assume that a large number of data have
already been gathered, and work on principles for selecting a subset of data
for efficient training. A drawback with the method is that the entire data set
has to be examined in order to decide which example to add to the current
training subset. Plutowski *et al* report, however, an order of magnitude faster
convergence than if the network was trained on the entire data set. Plutowski
et al use in the training process the least-mean-square-error function

$$E(w) = \frac{1}{2n} \sum_{p=1}^{n} \left(g(x_p) - f(x_p, w) \right)^2 \tag{6.85}$$

where $(x_p, g(x_p))$ is the pth input–output relation and $f(x_p, w)$ is the network
output on x_p.[10] A criterion for selecting training examples that works well in
conjunction with the error function used for training is the *integrated squared
bias* (ISB) given by

$$\text{ISB}(X_n) = \int \left(g(x) - f(x, w_n) \right)^2 \mu(dx) \tag{6.86}$$

where X_n is a data subset of size n, w_n is the set of weights that minimizes
(6.85) on X_n, and μ is a distribution over the inputs. Clearly, finding a
subset X_n that minimizes (6.86) gives us a subset representative of the whole
data set. Finding such a subset is computational impractical. Plutowski *et al*
approximate a solution by successively adding new examples x_{n+1} to the training
subset so as to maximize the decrement in ISB given by $\Delta\text{ISB}(x_{n+1}|X_n) = \text{ISB}(X_n) - \text{ISB}(X_n \cup \{x_{n+1}\})$. Using first-order Taylor expansions this decrement
can be approximated by

$$\Delta ISB(x_{n+1}|X_n) \approx \left(\begin{array}{c} (g(x_{n+1}) - f(x_{n+1}, w_n)) \\ f'(x_{n+1}, w_n)^T E''(X_n, w_n)^{-1} \end{array} \right)^T$$

$$\times \sum_{p=1}^{P} f'(x_p, w_n)(g(x_p) - f(x_p, w_n))$$

$$= \left(E'(x_{n+1}, w_n)^T E''(X_n, w_n)^{-1} \right) \sum_{p=1}^{P} E'(x_p, w_n)$$

$$\tag{6.87}$$

---

[10] We assume that the network has one output unit.

where the sum runs over patterns in the whole data set. This approximation depends on the weight update rule. Formula (6.86) is derived when using a Newton weight update rule. Because the mean least square error function is used, the Hessian can be approximated by $E'(X_n, w_n)E'(X_n, w_n)^T$. A similar rule could also be derived for gradient descent or conjugate gradient. These rules would be simpler and save computation, but not as accurate. It is possible to simplify (6.86) even more by ignoring all network gradient information. Interestingly, we then end up with a *maximum error* criterion selecting examples with maximum network error. This criterion is much cheaper to compute than (6.86) and works at least on some test problems in [199] as well as the original criterion. It is, however, not clear whether the network gradient information can be ignored in general. Plutowski *et al* also show that selecting examples by the ISB criterion works better than straightforward random sampling, which suggests that the validation scheme described in section 6.5.4 could be improved by exchanging the random sampling with a more sophisticated approach. Such an approach should, however, be independent of the size of the data set, so the ISB approach would need to be 'relaxed' somehow in order to be usable.

MacKay uses a different approach than Plutowski *et al* including noise in his model, but finally ends up with a similar result. He uses a Bayesian perspective to obtain an information based criterion about what example to pick next. The *posterior probability* of the weights $P(w|X, \aleph)$ can in Bayesian terms be expressed as

$$P(w|X, \aleph) = \frac{P(X|w, \aleph)P(w|\aleph)}{P(X|\aleph)} \tag{6.88}$$

where X is a subset of data and $\aleph$ is the network. The normalizing term $P(X|\aleph)$ is a constant and can be ignored in what follows. $P(X|w, \aleph)$ is often denoted the *sample likelihood* and is associated with the error of the data and $P(w|\aleph)$ is the *prior probability* of the weights, which is associated with regularization functions, such as weight decay [212]. The sample likelihood and the prior probability are defined as:

$$P(X|w, \aleph) = \exp(-\beta E_D(w)) \tag{6.89}$$
$$P(w|\aleph) = \exp(-\alpha E_w(w))$$

where β is the inverse of a noise parameter, $E_D(w)$ is an error function of the form (6.85), α is a regularization parameter, and $E_w(w)$ is a regularization function, such as weight decay. If we define $M(w)$ to be

$$M(w) = \beta E_D(w) + \alpha E_w(w) \tag{6.90}$$

then the posterior probability is given by

$$P(w|X, \aleph) = \exp(-M(w)). \tag{6.91}$$

Let X_n be a subset of examples of size n. Then $P(w|X_n, \aleph)$ is the posterior probability when these n examples are used in training. The idea now is to construct a measure of the information gained by adding a new example to the subset. Such a measure can be constructed by means of entropy functions. The entropy S_n of $P(w|X_n, \aleph)$ is defined as

$$S_n = -\int P(w|X_n, \aleph) \log P(w|X_n, \aleph) \, dw. \qquad (6.92)$$

The higher the value of S_n the more 'uncertainty' is associated with the weights. See for example [150] for further reading about entropy functions. The information gained by adding a new example x_{n+1} can now be expressed as the change in entropy $\Delta S_n = S_n - S_{n+1}$.

Let w_n denote the weights that minimize $M(w)$, i.e. maximize $P(w|X_n, \aleph)$, on the subset X_n. MacKay approximates the information gain by using a quadratic approximation of $M(w)$ expanded around w_n, and a first-order approximation of the new Hessian matrix $M''(w_{n+1})$ from $M''(w_n)$. Based on these somewhat crude approximations we finally get the formula

$$\Delta S_n \approx \tfrac{1}{2} \log \left(1 + \beta f'(x_{n+1}, w_n)^T M''(w_n)^{-1} f'(x_{n+1}, w_n)\right). \qquad (6.93)$$

If we compare this formula with formula (6.86) for ΔISB, we observe that there is a great similarity. The only main difference is that ΔISB involves error gradients of the new example while ΔS_n involves network gradients. Nevertheless, since the term $f'(x_{n+1}, w_n)^T M''(w_n)^{-1} f'(x_{n+1}, w_n)$ can be interpreted as the variance of the network when example x_{n+1} is sampled, we obtain the maximum information gain by picking the example with the largest error. This is consistent with the result of Plutowski *et al.* MacKay generalizes these results to selecting examples in specific regions of input space and selecting multiple examples. MacKay emphasizes, however, that some care should be taken when applying these techniques. The information gain estimates the utility of a data example assuming that the network model is correct, i.e. that the network is able to implement the desired and usually unknown input–output mapping. If the model is actually an approximation then the method might lead to undesirable results. For further reading about the Bayesian approach see [129], [179] and [180].

6.8   Conclusion

Training algorithms for feed-forward neural networks can roughly be divided up in two categories, gradient descent algorithms and second-order algorithms, such as conjugate gradient. There is up to date no simple answer to the question of which algorithm to prefer in general. Gradient descent exhibits linear convergence, but can, nevertheless, converge faster than second-order algorithms, when used in on-line mode and redundancy is present in the data. The author recognizes two approaches that should be explored further:

- Improve the on-line gradient descent techniques. The work of Darken *et al* and Le Cun *et al* described in sections 6.4.5 and 6.4.7 respectively are promising approaches in this area.
- Develop stochastic versions of standard second-order methods that can operate on subsets of data. The stochastic version of the scaled conjugate gradient algorithm described in section 6.5.4 is a promising example of such an approach.

One main conclusion is that *the* most effective general algorithm for training does not exist yet. What algorithm to use is still a problem-dependent matter. One has to consider whether the training has to be in *on-line* or *off-line* mode. On-line is usually the best for classification problems characterized by training sets containing a lot of redundant information. If on-line mode is used, the choice is between the stochastic scaled conjugate gradient algorithm described in section 6.5.4 and a carefully tuned on-line gradient descent algorithm combined with techniques like the reduction of large curvature components described in section 6.4.7. If off-line mode is used, then the choice is to use a second-order algorithm, and then the scaled conjugate gradient algorithm described in section 6.5.3 is a good choice.

Chapter 7

Exploiting Local Optima in Multiversion Neural Computing

Derek Partridge
University of Exeter, UK
derek@dcs.exeter.ac.uk

7.1 Local optima as a neural computing problem

The training of feedforward networks, such as multilayer perceptrons (MLPs) or radial basis function (RBF) nets, is well known to be highly sensitive to initial conditions. A change in the random initialization of the weights, even when all other conditions are fixed, will cause the learning process to converge on another minimum. The alternative minima on the error surface may result in drastically different generalization properties.

Sensitivity to weight seed variation has captured most attention, perhaps because it is somewhat unexpected. For if it is acceptable to initiate training from a random position in weight space, then it might be thought that the exact (random) position chosen will have no significant effect on the final outcome. However, for all but the most trivial problems the error surface is not as simple as the above expectation implies. Ridges and canyons in the error surface ensure that closely similar starting conditions can result in totally different trained networks.

In fact, the problem of convergence on different solutions becomes one of non-replicability of neural computing results. Given that the backpropagation learning process is one of iteration over the real numbers such that the final result is an accumulation of many small increments, *any* change in the actual representation of the real values in the hardware (which can be due to reading in weights from a stored file rather than from the random number generator as well as software and hardware updates) can alter the point of convergence on the error surface [75].

Curiously, systematic exploration of the sensitivity of training to initial conditions reveals that weight seed variation has the least effect, about the same as adding or deleting one hidden unit. As might be expected training set composition and structure (e.g. random versus boundary condition patterns) has considerably more effect, and changing the network type (e.g. MLP to RBF) has the most effect of all [74].

Sensitivity to initial conditions, in a general sense, has long been recognized and guarded against, usually by training a set of networks under a variety of conditions and then selecting the best (according to some test criterion), or taking some average measure as the expected performance level. This procedure considerably lowers the chances of settling on a poor local minimum but it involves, of course, an extra expenditure of resources, most of which are eventually wasted.

7.2   The group properties of local optima

There are two different views of this sensitivity issue, and both can be used to advantage. The first viewpoint is based on the fact that training neural networks is a process of approximation. A trained neural network is an approximate implementation of the function from which the input–output pairs, the relations that are the training set, have been drawn. Indeed, there are infinitely many such functions for which the trained net is an approximation. But usually we have a target function loosely defined (at least) and our concern is with the trained net as an approximation to this specific target function.

The case for this averaging strategy for sets of optima produced by means of random weight seed variation has been argued [58]. This argument hinges on the idea that extra complexity is introduced by the weight randomization. This leads to the conclusion that an infinite set of local optima (which will encompass all of the possible randomizations) would be an implementation of a less complex function (than any of the individual networks), and thus is expected to deliver the optimum performance. Any finite set of local optima generated with different weight randomizations is thus an approximation to this infinite optimum, and the optimal result is obtained by averaging across the set available.

A second viewpoint holds that different local minima will each compute correctly some input–output relations that the other local minima compute wrongly. Crudely stated, each of the different local minima will cover a different subset of the target function. This view leads to the idea that by proper organization and choice of the local minima to use on a given input, it should be possible to use the complete set of local minima collectively as a better implementation of the target function than any of the individual networks alone.

Several different strategies are immediately apparent:

- *Majority voting.* Under the assumption that the individual minima will each cover most of the target function similarly with only a minority making mistakes, a simple majority-vote strategy will deliver an optimal result for the set as a whole.

- *Expert selection.* Under the assumption that the differences between the individual minima are due to 'specialization' such that each individual network is correct on some part of the target function where the others are incorrect, selecting the right 'expert' network for a given class of input will deliver an optimal result for the system as a whole. Figure 7.1 is a schematic illustration of these two alternative multiversion system configurations.

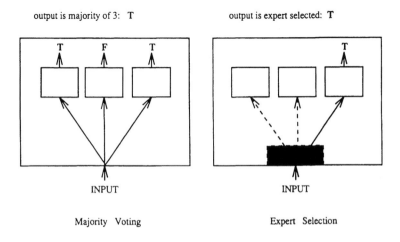

Figure 7.1. Two multiversion system architectures.

Clearly the efficacy of strategies rests on having sets of local optima with the necessary properties. For majority voting the desired property is lack of coincident failure among the set of local minima, the set of versions, i.e. if any one version fails on a given input the others should succeed. This property is known as *diversity*, but this term is mostly used, in a rather loose fashion, by software engineers in reference to alternative versions of conventional programs. Sometimes it denotes this lack of coincident failure property and other times it is used in a more general sense to denote any property of a set of versions that enables the set as a whole to be more reliable than any single version.

A formal definition of coincident-failure diversity has been provided [55], and extended to define a further measure, distinct-failure diversity. This is the property of different failures rather than simply success or failure, and is an overall measure of diversity that includes both types.

In general, a version set is maximally diverse if each failure is unique to just one version (in which case it must also be a distinct failure). It is non-diverse when all failures are duplicated in all versions (and again in this extreme case whether the failures are distinct or all identical is of no import)—i.e. the set of versions are duplicates of each other, and so there can be no gain from a consideration of the set rather than any one version.

7.3 Engineering productive sets of local optima

Having determined the properties that we wish a set of local optima to exhibit, the problem becomes one of constructing a set with the desired property. The averaging strategies exploit the reliability of large numbers. Assuming that all the networks are reasonably good approximations of the target function, then an average of many will, in general, be a more accurate result than the output of any individual network. To obtain such a set all that is required (after prototyping has revealed the range of usable hidden-unit numbers and training set sizes) is to systematically vary the initial conditions (within the constraints identified) and so produce a set of differing approximators.

For coincident failure diversity we have ranked the diversity-generating potential of the major parameters [75]. The strategy then is to vary these parameters giving most weight to the ones ranked highest. A further technique that has proved effective is to generate a 'space' of more trained networks than required, and then to select a maximally diverse subset heuristically.

A number of techniques for constructing sets of expert networks have been proposed, for example the gated experts of [70] which require that some initial decisions be made on how to split up the problem. The expert set is then constructed by using a 'gating' network to control the training of the experts in the set.

Another approach that we have explored uses a Kohonen feature map to classify the training set into some predetermined number of categories. Then the patterns in each category are used to train one expert network. The complete system requires that the Kohonen feature map preprocesses new inputs into one of the categories. This input is then passed on to the expert network which was trained on that particular category of training patterns, and this single, expert network computes the output. Figure 7.2 illustrates this particular multiversion software architecture.

Initial results on both well defined and data-defined problems indicate that this is a promising approach to the construction of sets of specialists in that it produces effective expert networks and requires no data-specific knowledge. However, an initial decision must be made on the number of categories to be used. Currently, this decision is somewhat *ad hoc*.

7.4 Optimizing performance over local optima sets

As neural networks produce real-valued results in their output units, averaging can be simply a task of summing these real values across all versions for each output unit, and computing the average value in each one.

OUTPUT

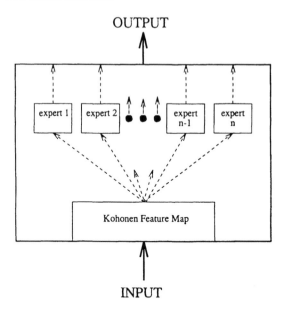

INPUT

Figure 7.2. An expert-net multiversion architecture.

However, as tasks become more complex other possibilities emerge. Consider a 1-out-of-N classification task, such as character recognition. A single version would likely have 26 output units (one for each possible result from A to Z). An ideal result would compute 1.0 in the correct output unit and 0.0 in the other 25. But an actual result is likely to be a value between 0.0 to 1.0 (inclusive) in *each* of the 26 output units. This would occur in every one of the versions in the system.

The problem is how best to average these results. Take the maximum from each version (as it is only supposed to identify one letter for each input), and compute and identify which class has the highest average over all versions? Or simply average all values in all classes over all versions and take the highest as the result? Or average all values exceeding a threshold (say, 0.5) in all classes over all versions, and take the highest as the result?

All three of these options are quite reasonable. One factor that will influence their relative merits will be the training regime. How was a correct result for training purposes computed?

We have explored these three averaging strategies on a letter recognition task [76]. We found that in general the best results were produced when all output values were carried forward into the final averaging process (i.e. not maximum only and not arbitrary thresholding).

Coincident-failure diversity is exploited using some majority-vote strategy. The simplest approach is to take the majority result, of the set of versions, as the overall result. But in cases where correctness is particularly crucial then a system result may not be acceptable unless all versions agree.

Notice that with a majority-vote strategy an optimum system (i.e. 100% correct) may be obtained from a less than maximally diverse system. As long as the coincidence of common error is restricted to a minimum of versions, then a majority outcome will always be correct.

Once distinct-failure diversity is taken into account optimum system performance is obtainable with even more error in the overall system. Consider for example a classification problem with m different output categories, e.g. character recognition where the input image is to be classified as one of the 26 upper-case characters, so $m = 26$. In this example, for any given input one output class is correct and there are 25 distinct ways of generating an incorrect output, i.e. there are 25 distinct ways for the network to fail.

If, in a set of N networks, just two compute the correct output class and $N - 2$ versions fail but fail distinctly, i.e. no two failures in the same output class, which implies $(N - 2) \leq 25$ and $N > 2$, then a majority in agreement strategy will deliver a correct answer. So even when most versions incorrectly compute all inputs, a properly organized system can perform 100% correctly!

Exploitation of sets of expert networks requires some decision strategy to preprocess inputs and determine which of the expert networks should compute each particular input. In the 'gated' expert sets, mentioned above, as part of the training process the gating network learns the probability to associate with each expert network according to the characteristics of each input. The system output may, in general, be a weighted mix of the individual expert net outputs. In the limiting case of probabilities of zero and one, the function of the gating network will be to 'turn off' all the expert networks except one which is 'turned on' with an associated probability of one.

Using the Kohonen feature map approach, it is the particular map learned during initial classification of the training data that is used as the multi-way switch for the functioning system. Again in the extreme case of each input being assigned to just one class, the particular class indicates the single expert network to be used. However, there is also scope for using more subtle mixes of experts.

Such a feature map actually computes a 'distance' of each new input from each of the classes. So this distance measure may be used to provide a probability that each expert will correctly compute any given input. Taking such a course, each new input may then be computed by all experts with a non-zero probability of being correct, and the final system outcome obtained from this probabilistically weighted subset of networks.

7.5 Applications of the multiversion approach

Both approaches to the exploitation of sets of local minima have proved to be effective on a variety of problems from implementation of well defined abstract functions, to data-defined problems such as character recognition and vowel discrimination tasks.

An averaging approach has been favoured by the neural computing community when dealing with non-expert sets of alternative versions.

Strategies for exploiting minimum coincident failure among a set of alternative versions arose in the context of software engineering efforts to construct ultra-reliable systems. The alternative versions are alternative programs (implemented, say, by different programmers or in different languages), and the general idea is called *multiversion* or *N-version* software engineering [60].

Application of these ideas within neural computing opens several new options. Firstly, obtaining high levels of the necessary diversity in version sets is most easily done by using diverse programming methodologies (e.g. Prolog versions and Modula-2 versions). Neural computing then offers the software engineer a radically diverse methodology in comparison to any conventional programming language [73]. So further improvements in system reliability (for suitable problems) may be obtained by adding neural-net versions to the conventionally programmed versions to obtain the complete multiversion system.

Secondly, the unique nature of neural computing in comparison with conventional programming makes multiversion software engineering considered as a purely neural computing activity a more attractive prospect.

- Automatic training is so much faster and cheaper than conventional programming that the economics of a multiversion strategy become much more favourable.
- The characteristics of a trained network (i.e. the precise local optimum reached) are wholly determined by the initial conditions for training, and are thus amenable to manipulation in order to systematically generate high diversity. In conventional programming the process cannot be manipulated with precision (it is a manual art).

In sum, it is relatively cheaper and relatively more feasible to engineer the required diversity into sets of neural networks than into sets of conventional programs.

7.6   Discussion and conclusions

Although the two different strategies for exploiting sets of local optima have been presented, and sub-strategies also described, it is quite possible to view all of these options as points along a continuum.

At one end, carrying all raw output results forward for a final averaging is a strategy in which all information generated by the networks is fed into the final outcome. As this has generally proved to be an optimal averaging strategy, we must conclude that the incorrect positive values carried forward are more than compensated for by the correct, but non-maximum, results which are also carried forward.

At the other end of the spectrum, for a set of non-overlapping expert networks just one specific version is selected (and just the maximum output of that version) to determine the final system outcome. In between these two extremes, minimum coincident failure sets contribute a majority of version results to determine the system outcome.

As experts begin to share more of the target function relations in common then they become more like minimum coincident failure sets. As 'averaging' sets are treated by just considering the maximum value output from each version so they also become more like minimum coincident failure sets.

At the extremes these options are clearly distinct, and so given a set possessing one of the extreme characteristics it is clear which strategy will best exploit it. But as soon as less extreme sets are available (which is, of course, the situation in practice), then the decision on an optimal strategy is no longer clear-cut. Thus there is considerable scope for research on most (if not all) of the issues raised above.

In particular, consideration of the interaction between the details of the training regime and the choice of an optimal version-set exploitation strategy has received little attention. At the outset, which general multiversion approach to pursue given a specific problem, the details of the data available and the requirements of the final implementation are far from well understood.

Acknowledgments

Niall Griffith first proposed and explored the Kohonen feature map approach to expert networks. The support of a visiting professorship from the Universiti Sains Malaysia, Penang is gratefully acknowledged as well as longer-term support for the prerequisite research from the Safety-Critical Systems programme of the DTI and EPSRC in the UK (grant No GR/H85427).

PART 3

APPLICATIONS

The number of commercial and industrial uses of neural networks is increasing rapidly. Their use encompasses fields as diverse as healthcare, marketing and control of industrial processes. It would be impossible to give a comprehensive overview of all neural networks applications in one book. However, the following part of this book contains three interesting and diverse applications.

Chapter 8

Neural and Neuro-Fuzzy Control Systems

Phil Picton
Nene College, UK
phil.picton@nene.ac.uk

8.1 Introduction

This chapter tries to explain how neural networks are being used in control systems. This is still very much a research area and most work is being done on simulations rather than on real applications. However, neural networks, either on their own or in conjunction with fuzzy systems, appear to have great potential to control systems where traditional control has failed, in particular, in the automation of systems where, at present, the only way of controlling the system is to use humans. Neural networks and fuzzy logic are able to emulate the way in which humans appear to think, and therefore provide a way of substituting the human with an automatic controller.

Since this is an area of active research, established procedures have not yet been developed. In this chapter, the basic ideas that are the starting points for much of the research are given. The aims of the research, and the basic ideas behind them should give a good grounding for anyone who wishes to pursue these ideas further.

8.2 Traditional control

There is no denying that control theory, or at least linear control theory, has been very successful. Systems can be controlled by comparing the value of the output with the desired value, and the error signal used to drive the system towards the system goal. Figure 8.1 shows a block diagram of such a control system.

As you can see, the connection from the output to the comparator (the feedback path) has no box associated with it, which means that its transfer function is 1 or unity. In practice, the feedback path may not be unity, but any

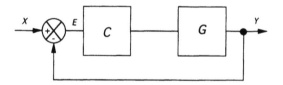

Figure 8.1. Linear control system with unity feedback.

system can be rearranged to a unity feedback system for the purpose of analysis. The path between the comparator and the output (known as the forward path) has two boxes: the controller with a transfer function C, and the system itself with the transfer function G. The output, Y, is defined by equations (8.1) and (8.2):

$$Y = CGE \tag{8.1}$$

$$E = X - Y. \tag{8.2}$$

If the system is linear, then these equations can be combined to give the transfer function of the whole system,

$$\frac{\text{output}}{\text{input}} = \frac{CG}{1 + CG}. \tag{8.3}$$

If the system is non-linear, then it is usually impossible to derive a transfer function for the whole system. The transfer function comes in two forms, depending on the type of system. If any part of the system uses sampled data, such as when the controller is a computer, then the transfer function uses the z-transform. If no part of the system is sampled, then the Laplace transform or s-transform is used. These transfer functions are models of the behaviour of the parts of the system. They would have been calculated from data collected about the system, and are approximations to the physical system.

Linear models have proved to be so useful that they are used as often as possible, even to the point of 'linearizing' the physical system so that it behaves more like a linear model. However, there are bound to be some systems which are clearly not linear, and refuse to be manipulated in order to appear linear. In these circumstances, more sophisticated techniques will be used. Although these more sophisticated methods are commonplace in the literature and are taught in control engineering courses, in general they do not have the same appeal as linear controllers, and are certainly less robust.

There is inevitably a debate between control theorists as to which is the best approach to controlling systems for which no suitable linear model can be found. The reasons might be non-linearities in the system, a time-varying system, or simply one which is not well understood for which there is no known model. One approach is to use self-tuning or adaptive controllers. These have been used and are quite robust. However, understanding how they work, and guaranteeing their robustness and reliability is not usually possible. The average control engineer is therefore stuck, without a way forward in these circumstances.

8.3 State space

Another way of representing a system is to use state space notation. In state space, a system is broken down into a set of first-order variables or states. The advantage of the state space method is usually that it becomes easier for computer manipulation than transfer functions. However, it also brings out the idea that in order to control a system you need to know the values not only of the variables but also of their higher derivatives.

As an example, a system is modelled as a second-order differential equation of the form

$$\ddot{y} + 2\dot{y} + 3y = 4u. \tag{8.4}$$

The states can be found as follows in equations (8.5) to (8.8):

$$x_1 = y \tag{8.5}$$

$$x_2 = \dot{y} = \dot{x}_1 \tag{8.6}$$

$$\dot{x}_1 = x_2 \tag{8.7}$$

$$\dot{x}_2 = -3x_1 - 2x_2 + 4u. \tag{8.8}$$

In matrix form, these equations become:

$$\left| \begin{array}{c} \dot{x}_1 \\ \dot{x}_2 \end{array} \right| = \left| \begin{array}{cc} 0 & 1 \\ -3 & -2 \end{array} \right| \left| \begin{array}{c} x_1 \\ x_2 \end{array} \right| + \left| \begin{array}{c} 0 \\ 4 \end{array} \right| u \tag{8.9}$$

$$y = \left| \begin{array}{cc} 1 & 0 \end{array} \right|. \tag{8.10}$$

It can be seen that each of the state variables, x, can be expressed as a first-order differential equation. In general, any linear system can be summarized by

$$\dot{x} = Fx + Gu \tag{8.11}$$

$$y = Hx. \tag{8.12}$$

In the following sections, whenever the inputs to the controller are referred to this is meant to be inferred as the state inputs which contain the derivatives of the system inputs.

8.4   Neural control

Probably the best way to describe the work done on neural networks in control is to say that it has progressed in parallel with research in adaptive control. Many of the pioneers of adaptive control are now to be found working in the area of neural networks, so clearly there is a significant overlap. The difference between them, I suspect, is in the mathematical representation of the two models. Adaptive control theory uses complex mathematics. Although the same could be said of neural networks, in practice the mathematics can to some extent be avoided. It is therefore easier to picture and understand what is happening in a neural network than it is in an adaptive controller.

8.4.1   Approximating the inverse transfer function

The holy grail in control engineering is to be able to find the inverse model of the system that you are trying to control, or at least to be able to approximate it. Figure 8.2 shows an open-loop controller.

Figure 8.2. Open-loop controller.

If the controller, C, equals the inverse of G, then their product is 1, and the output follows the input. In feedback control, the transfer function of the whole system (given above) shows that if the controller has a constant gain, k, then increasing the value of k means that the transfer function approaches 1. This was what people originally did, and were surprised that this sometimes caused the system to become unstable. We can see what is happening in a closed-loop system if we rearrange the diagram so that closed-loop control becomes open loop, as shown in figure 8.3.

The feedback path in the controller contains the transfer function, G, which is the system under control. If a controller was constructed this way, then this transfer function has to be approximated. The approximation is given the symbol $\hat{G}$. The controller transfer function (taking the whole of the dashed box as the controller) is shown in equation (8.13):

$$\frac{k}{1 + k\hat{G}}. \tag{8.13}$$

If the value of k is increased, this approximates to $1/\hat{G}$ or $\hat{G}^{-1}$, the inverse transfer function of the system being controlled.

The goal is therefore to emulate the inverse transfer function of the system under control. Since neural networks have the ability to emulate any mapping

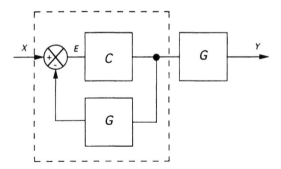

Figure 8.3. Rearranged closed-loop system.

between two variables, the inverse transfer function can be found, as described
in [65]. In particular, a multi-layered perceptron would be used, with the inputs
to the system as the desired outputs of the network, and the outputs of the system
as the given inputs, as shown in figure 8.4.

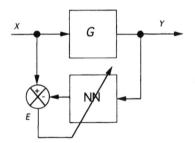

Figure 8.4. MLP emulation of $\hat{G}^{-1}$.

Hornick [46] showed that a three-layered multi-layered perceptron can
emulate any function. Two problems still remain, however. The first is that
the number of units in the hidden layer could be infinite. The second is that
no proof yet exists that shows that back-propagation can guarantee to find the
solution. It is generally accepted, even without these two problems, that a
network can be found and that back-propagation will find a solution. Often
more than three layers are used, which can speed up the learning process.

8.4.2 Emulating a controller

One of the earliest examples of a neural network in a controller was the
ADALINE, described in [105], which could learn to behave in the same way as
a human controller. Unlike the previous section, in this work the aim is to learn
the transfer function of the controller, which is a human. The ADALINE is
therefore connected so that the inputs to the controller are the same as the inputs

to the ADALINE, and the outputs of the controller are the desired outputs of the ADALINE. There are many limitations to this approach, not least of which is that the ADALINE is a single-layered network which means that only some relationships between input and output can be learnt. An alternative is to again use a multi-layered perceptron in place of the ADALINE. This would overcome this limitation and allow any controller to be emulated.

8.4.3 Reinforcement learning

Reinforcement learning is a way of training a system to avoid failure. In control problems where the goal is to keep a system within certain parameters, if the system moves outside of those parameters then it is said to have failed.

Any dynamic system can be represented in a state space (described earlier), which shows the state of the system at any given time in terms of the control parameters and their derivatives.   Donald Michie at Edinburgh University produced an early attempt at controlling a system which divided the state space up into boxes, hence the name of this system, BOXES. The state space was divided up and within each box a pair of weights would be adjusted until the system was satisfactorily controllable.

The learning mechanism is as follows. Initially, each box has two weights— one which favours a positive outcome and the other a negative outcome. This means, in control problems, that the controller has 'bang–bang' control, where the driving force for the system is either maximum positive or maximum negative. This is clearly a severe restriction on the control system, but one which has some advantages. Pontryagin's maximum principle states that time-optimal control can be achieved using bang–bang control. However, this has to be treated with some caution since in many circumstances time-optimal control is not necessarily the best option.

The weights are initialized to small random values. In any particular box there will be two weights, w_p and w_n, which correspond to a positive or a negative decision respectively. When the box is first entered, the resulting output is determined by which of the two weights is the larger. So, for example, if w_n is larger than w_p, the resulting output is negative. While the system remains within the parameter boundaries a count is kept. When the system fails, the count is noted and the weights adjusted so that they represent the average time to failure.

The BOXES system could be interpreted as a rule-based controller which has the ability to learn. However, the rules which divide up the space do not adapt, so there has to be a careful decision made by the designer at the start. Some attempts, such as that by Woodcock [115], have been made to overcome this limitation by making use of genetic algorithms (GAs). The GA evolves a system of boxes which contain a control decision. Over time, the GA evolves a system which is controllable. A variation on this was the use of 'fuzzy' boxes, where the borders of the boxes were deliberately fuzzified. This had the effect of speeding up learning, but if the boxes remained fuzzy the final performance was poor. This suggests that a system which is initially fuzzy, but which becomes more crisp over time would be a good solution. These ideas have yet to be tested on anything other than simple simulations.

The adaptive critic element (ACE) and associative search element (ASE) of Barto [6] was inspired by the BOXES approach, but took it a stage further by being able to predict failure rather than having to wait for it to happen. However, the initial decisions on how to divide up the state space are the same as in BOXES and therefore subject to the same criticism.

Figure 8.5 shows a schematic diagram of the ASE and ACE. The state variables from the system being controlled are decoded into binary variables, so that when the system is in the equivalent of one of Michie's boxes, only one of the variables x_i is 1. The ASE is like a neuron which produces a weighted sum of the inputs. Since only one input is 1, the weighted sum equals the weight on the active input. However, random noise is added so that, when the system starts, the weights are zero but the sum has an equal probability of being positive or negative.

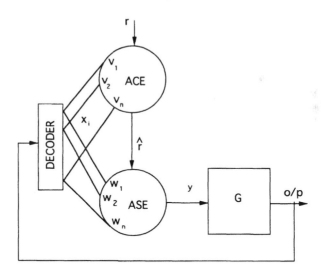

Figure 8.5. ASE and ACE system.

The single weight acts like the difference between the two weights, w_n and w_p in Michie's BOXES, however, the weight adjusting mechanism is different. An earlier version of the system only consisted of the ASE, and the reinforcement signal, $r(t)$ was applied directly to the ASE. The weights were adjusted using equation (8.14) where α is a small positive constant, $r(t)$ is the reinforcement at time t, and $e(t)$ is the eligibility at time t:

$$w_i(t + 1) = w_i(t) + \alpha r(t)e(t). \tag{8.14}$$

The reinforcement, $r(t)$, has a value of 0 during training and changes to -1 upon failure. The eligibility is a function given by equation (8.15) where δ lies between 0 and 1:

$$e_i(t+1) = \delta e_i(t) + (1 - \delta)y(t)x(t). \qquad (8.15)$$

The eligibility performs a similar function to the count in Michie's BOXES. When a box is entered, the eligibility for that box starts to rise exponentially, and when the system leaves the box, the eligibility drops exponentially. The longer the system stays in a box, the larger the eligibility for that box, but the longer the time between leaving the box and eventually failing, the smaller the value of the eligibility for that box. Boxes with high eligibility correspond to boxes that the system was in just before failure.

When a particular box is entered, x_i for that box is 1 while all other x are 0. An output decision is made, so y is either $+1$ or -1. Initially, $e_i(0) = 0 =$, so $e_i(1) = (1 - \delta)y(0)$, which is either $+(1 - \delta)$ or $-(1 - \delta)$. For example, if $y = +1$ and $\delta = 0.8$, then $e_i(1) = 0.2$. The next time the system is sampled, if it is still in the same box, the eligibility is given by

$$e_i(2) = \delta e_i(1) + (1 - \delta) = 0.8 \times 0.2 + 0.2 = 0.36. \qquad (8.16)$$

Alternatively, if a new box is entered, x_i becomes 0, and the eligibility drops to that given by

$$e_i(2) = \delta e_i(1) = 0.8 \times 0.2 = 0.16. \qquad (8.17)$$

This continues, with $e_i(t)$ getting smaller and smaller. Looking back at the equation for the weight adjustment, it is possible for the value of $e_i(t)$ to be substituted but, until the system fails, the value of $r(t)$ is 0, so the weight stays the same. When the system fails, $r(t) = -1$, and $e_i(t)$ has dropped to a small value. The amount that is subtracted from the weight depends on the length of time before failure.

The ASE on its own can only ever adjust the weights upon the system failing. In order to also adjust the weights continuously the ACE is used. In this, a prediction signal, $p(t)$, is produced, based on the current set of weights and input values,

$$p(t) = \sum_{i=1}^{n} v_i(t)x_i(t). \qquad (8.18)$$

This is a prediction of the final amount that will be deducted from the weight when the system fails. It produces an estimate of the reinforcement signal, $\hat{r}(t)$, as its output, shown in equation (8.19), where γ is between 0 and 1:

$$\hat{r}(t) = r(t) + \gamma p(t) - p(t - 1). \qquad (8.19)$$

The weights are continuously being adjusted using

$$v_i(t+1) = v_i(t) + \beta \hat{r}(t)q_i(t). \qquad (8.20)$$

The variable $q_i(t)$ is called the trace and plays the same role as the eligibility in the ASE, the difference being that it is independent of the value of the output. It is found using equation (8.21) where λ is between 0 and 1

$$q_i(t+1) = \lambda q_i(t) + (1 - \lambda)x_i(t). \qquad (8.21)$$

The weights stop changing when $\hat{r}(t) = 0$, which happens when $\gamma p(t) - p(t-1) = 0$. Usually, γ is made to be close to 1 (0.95 say), so that the weights stop changing when the predictions are roughly the same.

This method allows a system with no model available to be controlled using only a failure signal. The ASE provides the control signal, while the ACE evaluates the state of the system. This is a very useful control strategy for certain types of system.

8.5 Fuzzy control

8.5.1 Background

Fuzzy logic has made a great impact in control engineering in the last decade. The initial ideas proposed by Lotfi Zadeh were rather mathematical and were published in relatively obscure conference proceedings and journals. However, the Japanese took these ideas and started to use them in a more practical way, with the result that fuzzy control is making inroads into engineering.

8.5.2 Fuzzy logic

The ideas involved in fuzzy logic allow us to combine in a 'logical' way some weighting factors associated with propositions from different sources. Propositional logic uses the logical operators AND, OR and NOT to combine logical inputs. In fuzzy logic we have equivalent operations, namely MIN, MAX and $(1 - x)$ respectively.

When the truth values of 0 for false and 1 for true are used, there is no difference between Boolean logic and fuzzy logic. Fuzzy logic, however, can generalize so that the truth values can be any number between 0 and 1.

8.5.3 Fuzzy sets and membership functions

Fuzzy logic enables you to combine logical values together. However, using the operations of MIN, MAX and $(1 - x)$ it can combine real numbers between 0 and 1 that are called the fuzzy set memberships. These numbers indicate the level of membership of a set, where a set can be a description or quality of an object. This is probably easier to explain with an example.

Figure 8.6 shows how some arbitrary thresholds, T_1 and T_2, could be set along a temperature scale. At any temperature, T, the temperature is either COLD when $T \le T_1$, HOT when $T > T_2$, or WARM in between. Therefore, all

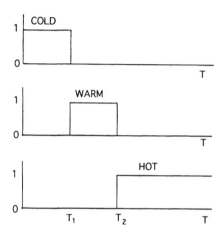

Figure 8.6. Temperature scale with thresholds.

values of temperature which are less than T_1 belong to a set which is labelled COLD.

Membership of a set is represented by a value of 1, whereas non-membership is represented by a value of 0. Any temperature T can only be a member of one set. In other words, the set called COLD contains all temperatures which are less than or equal to T_1. These membership functions are shown in figure 8.6. The membership can be used as the truth value so that, for example, if the membership of the set COLD is 1, then this is equivalent to the proposition (temperature is COLD) being TRUE. At the same time, membership of the other two sets is 0 and is equivalent to the propositions (temperature is WARM) and (temperature is HOT) being FALSE.

The boundary between two sets is very 'crisp'. For example, if T_1 is 10°, a temperature of 5° is clearly COLD. However, at a temperature of 9.999° it is not so obvious whether this should be interpreted as COLD or WARM—it is somehow both.

In fuzzy logic, the shape of the membership functions is more flexible so that the boundary does not have to be crisp. Figure 8.7 shows an example, using the commonly found triangle functions.

The most obvious difference is that the sets overlap, so that at some temperatures it is possible to be a member of two different sets to some degree. At the temperature shown in figure 8.7, the memberships are VERY COLD (0.7), COLD (0.3), WARM (0.0) and HOT (0.0).

8.5.4 Fuzzy logic control (FLC)

One way of describing fuzzy control is to think of it as an improvement on rule-based control. In rule-based control, the controller consists of a set of rules

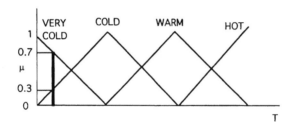

Figure 8.7. Fuzzy sets.

which quite often have been elicited from human controllers. When applied to a system, it is often found that the system response is not very smooth—far worse than the human controller. Fuzzy logic control can start with the same set of rules, but smooth the operation of the system by making the inputs and outputs fuzzy.

For example, a rule-based temperature controller could have rules of the form:

IF (temperature is VERY COLD) THEN (turn heating to HIGH)
IF (temperature is COLD) THEN (turn heating to MEDIUM)
IF (temperature is WARM) THEN (turn heating to LOW)
IF (temperature is HOT) THEN (turn heating to OFF)

When these rules are applied, terms like HIGH, MEDIUM, LOW and OFF have to be defined. From what was said earlier about crisp sets, it is not difficult to see why a rule-based controller produces a response which is not smooth. As the temperature changes, a different rule will start to fire, causing an abrupt change in the system output. Note that in a rule-based system, only one rule fires at a time. The rule is selected as the one whose antecedents are satisfied. When fuzzy variables are used, this smooths the operation of the controller. The main reason for this is that all of the rules fire, and their actions are combined.

Using these rules, if the membership of VERY COLD is 0.7 at a particular temperature, then the heating is turned on HIGH with a membership of 0.7. In other words, the membership value is passed on to the action part of the rule.

At the same temperature the membership of the set COLD is 0.3, so the second rule turns the heating on to MEDIUM with a membership of 0.3. Figure 8.8 shows the membership functions for the heater which are also triangular, with the operating range of the heater from 0 to 15.

The membership values that are passed to the output appear as the shaded areas on the diagram. Defuzzification is used to calculate the final setting for the heater which finds the 'centre of gravity' of the shaded area. This is shown on the diagram, and is the point at which the shaded area to the left of the point equals the shaded area to the right. This point turns out to be approximately 12,

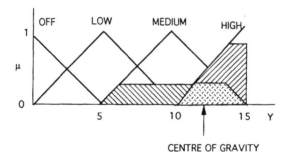

Figure 8.8. Membership functions of heater.

so the heater would be turned up to a setting of 12.

The formula for finding the centre of area is given by equation (8.22), where i is the number of membership functions and $\mu_i(y)$ is the membership value of the ith membership function at the point y:

$$y = \frac{\sum_{i=0}^{i=4} \int_0^{15} \mu_i(y) y}{\sum_{i=0}^{i=4} \int_0^{15} \mu_i(y)}. \tag{8.22}$$

This is quite a time-consuming equation to calculate each time. However, if we make certain stipulations such as the membership functions are symmetrical, and at any point the sum of the membership values is 1, then it is possible to replace each of the membership functions by a singleton, positioned at the centre of the triangular membership function, and given the height of the shaded area. Further, if we assume that the two triangular functions at the extremes are made symmetrical, then the formula for finding the centre of gravity is given by equation (8.23), where y_i^c is the centre of the triangular membership functions, or the position of the singleton:

$$y = \sum_{i=0}^{i=4} \mu_i y_i^c. \tag{8.23}$$

The advantage of using fuzzy rules is that the models and measurements do not have to be precise. The designer can use loosely defined terms like large, medium and small, and the membership functions themselves can be defined very loosely. However, this does require a certain amount of good guessing.

Research in this area has now focused on setting some of the parameters such as finding the membership functions of the fuzzy sets, using adaptive methods such as neural networks. Firstly, a rough guess is made of the fuzzy rules. Then the fuzzy rules are transformed into an equivalent neural network. The network is then shown examples and the weights adjusted to improve the overall performance. Then the network is transformed back into a fuzzy rule-based system.

The advantage of the neural network is that it provides an adjusting mechanism, whereas the advantage of the fuzzy rule-based system is that it can be efficiently coded and is robust. The following section describes some of the ways that neural networks and fuzzy logic can be combined.

8.6   Neuro-fuzzy control

Neuro-fuzzy control is a means of allowing a fuzzy control system to adapt to changes in the environment or to optimize the system after starting with an initial guess. Although systems now exist which use neuro-fuzzy control, it is still essentially a new area of research, and therefore there are no standard ways of implementing it. In this section three ways are examined. Each of these methods have features in common, but differ in detail. Essentially they all start with a fixed set of membership functions. Then either rules are generated by the system, or existing rules are weighted or given a confidence factor. This weight is what usually makes a neuro-fuzzy system adaptive.

8.6.1   Adaptive fuzzy associative memory (AFAM)

The fuzzy associative memory (FAM) by Kosko [54], is a method of interpreting rules as a mapping between the input membership values and the output membership values. One way of viewing the operation of a FAM is to map the existing rules and membership functions onto a space. A simple example is a heating controller which has the temperature as the input and the setting on a heater/cooler as the output. The rules that govern the fuzzy logic controller are:

IF (temperature is VERY COLD) THEN (turn heating on HIGH)
IF (temperature is COLD) THEN (turn heating on MEDIUM)
IF (temperature is WARM) THEN (turn heating on LOW)
IF (temperature is HOT) THEN (turn heating on OFF).

The membership functions for the input and output are shown in figure 8.9. The membership functions are drawn such that the inputs are on the y-axis and the outputs are on the x-axis, then areas are mapped out on the space when the rules are applied. Note that the areas overlap, so that a point in the space could belong to more than one area.

By looking at the operation of an existing (probably human) controller, the rules can be modified so that the mapping is improved. A system that does this is called an adaptive FAM or AFAM. An existing controller is used to control the temperature. Input/output pairs are generated, which can also be plotted in the space, which generally form clusters. The aim of the neuro-fuzzy controller is to adjust the rules so that they encompass the clusters. More specifically, points are found which correspond to representative members of each cluster, and these representatives are used to position the membership functions.

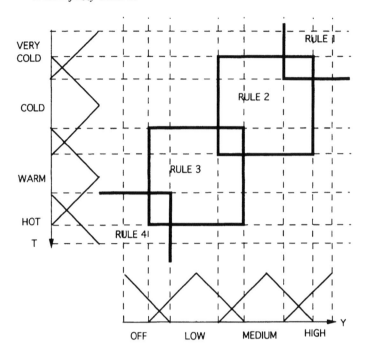

Figure 8.9. FAM.

One way of finding representative members of a cluster is to use learning vector quantization (LVQ), such as the network described by Kohonen [53]. The network consists of neurons whose weights are adjusted so that each neuron corresponds to a vector that is positioned in the centre of the cluster. Figure 8.10 shows an LVQ network.

For neuro-fuzzy control, the inputs to the LVQ network would be the pairs of points generated by the inputs to the control system and the corresponding controller outputs. These neurons in the LVQ network have weights which are initially randomized, so that each neuron can be regarded as a vector which has one end at the origin, and the other end defined using the weights as co-ordinates. The inputs are fed to each of the neurons in the network, so that each neuron receives the same inputs. The output of each neuron is the Euclidean distance between the input vector and the weight vector, given by

$$d_i = \left\{ \sum_{j=1}^{N} (w_{ij} - x_j)^2 \right\}^{\frac{1}{2}}. \tag{8.24}$$

The winning neuron, i.e. the one nearest the input vector which consequently has the lowest output value, has its weights adjusted so that the

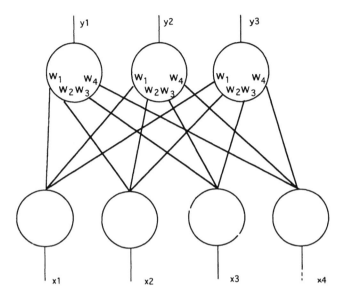

Figure 8.10. LVQ network.

corresponding weight vector moves closer to the incoming vector, using

$$\acute{w}_{ij} = w_{ij} + \alpha(x_j - w_{ij}). \qquad (8.25)$$

After training, the weights in each of the neurons will have been modified so that now each neuron corresponds to a vector that is inside a cluster of training points. Sometimes there will be more than one vector per cluster, which for fuzzy control is acceptable.

Inevitably, there will be a small number of points which lie outside of any of the areas covered by the fuzzy rules. By finding representative vectors, a more general view of the data is found which corresponds to the major features. Thus the rules are adjusted so that the points found by LVQ lie inside areas.

The question of exactly how you then adjust the rules to best fit the representative points is an area of active research. Bart Kosko suggests that the rules are weighted, and the number of points enclosed by the area is used as an estimate of the weight. Alternatively, the number of representative points within each area is used to weight the rule. If cells are overlapping, then the representative point is used to boost the weight of the area whose centroid is closest.

This method of finding clusters firstly gives an indication of the rules that should be used, i.e. wherever there is a cluster of points there is a corresponding rule. Secondly, the weights can then be updated as new data are collected, which gives the system the ability to adapt to a changing environment.

8.6.2   Takagi–Sugeno–Kang (TSK)

Another method which has been used as a basis for much research in neuro-fuzzy control is the Takagi–Sugeno–Kang method (TSK) from Takagi and Sang [94] and Sugeno and Kang [93]. In this method, fuzzification is as usual, but the rules are different. Typically a set of rules has the form:

$$Rule\ 1: \quad IF\ x_1\ is\ X_{1.1}\ and\ \ldots\ x_p\ is\ X_{p.1}$$
$$THEN\ y_1 = b_{0.1} + b_{0.1} + b_{1.1}x_1 + \cdots + b_{p.1}x_p$$

and

$$Rule\ i: \quad IF\ x_1\ is\ X_{1.i}\ and\ \ldots\ x_p\ is\ X_{p.i}$$
$$THEN\ y_i = b_{0.1} + b_{0.i} + b_{1.i}x_1 + \cdots + b_{p.i}x_p$$

and

$$Rule\ k: \quad IF\ x_1\ is\ X_{1.k}\ and\ \ldots\ x_p\ is\ X_{p.k}$$
$$THEN\ y_k = b_{0.1} + b_{0.k} + b_{1.k}x_1 + \cdots + b_{p.k}x_p.$$

The terms $X_{p.i}$ are linguistic fuzzy terms. The consequent of each rule is a linear weighted sum of the input values. The final output of the controller, y, is calculated using equation (8.26), where the weight h_i is the overall truth value for the premise of rule i:

$$y = \frac{\sum_{i=1}^{k} h_i y_i}{\sum_{i=1}^{k} h_i}. \tag{8.26}$$

The value of h_i is calculated using

$$h_i = \mu x_{1.i} \wedge \cdots \wedge \mu x_{p.i}. \tag{8.27}$$

The output is already a real-valued number, and therefore defuzzification is unnecessary.

In order to understand this method more clearly, consider the case of a single variable function, such as that shown in figure 8.11.

The membership functions are selected as three overlapping triangles. The width of each triangle defines a range over which the function is approximated. As shown in figure 8.11, the approximations are linear. We therefore have three rules. In each rule, since there is only one variable, x, some of the subscripts have been dropped in the antecedent parts.

$$Rule\ 1: IF\ x\ is\ X_1\ THEN\ y_1 = b_{0.1} + b_{1.1}x$$

$$Rule\ 2: IF\ x\ is\ X_2\ THEN\ y_2 = b_{0.2} + b_{1.2}x$$

$$Rule\ 3: IF\ x\ is\ X_3\ THEN\ y_3 = b_{0.3} + b_{1.3}x$$

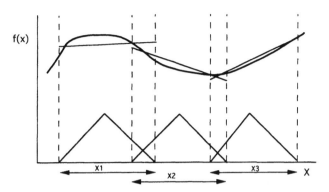

Figure 8.11. Single-variable function.

As you can see, the x-axis has been divided into X_1, X_2 and X_3 which overlap. In each of these divisions, y_i is a linear function of x_i. The final output, y, is found using equation (8.28), in which the weight h_i is taken as equal to μx_i

$$y = \frac{\sum_{i=1}^{3} h_i y_i}{\sum_{i=1}^{3} h_i}. \tag{8.28}$$

Johansen [48] showed that if the error function is taken as the mean squared error which is then multiplied by the membership value, then this combination of the individual linear approximations is optimal. The TSK method therefore has a means of fitting rules to the data. This means that it can find the initial rules, and has the ability to update the rules if new data arrive.

8.6.3   B-spline functions

A method which could be considered as a combination of the previous two, is that of Brown and Harris [7] who use b-splines to define the membership functions. The goal of their method is to produce a set of rules with confidence measures. However, they arrive at these rules and confidences by finding a set of weights which, when multiplied by the membership values of the input variables, can be converted into the rules and confidences.

The membership functions are produced using b-splines, with the order of the b-splines selected in advance. Figure 8.12 shows the membership functions produced when the order, k, is chosen as 1, 2 and 3.

When k is 1, the membership functions define a crisp set, so the function is represented as a constant. When k is 2, the b-splines are linear and the membership functions are the familiar triangles. When k is 3, the b-splines are quadratic functions, which produce curved bell-shaped membership functions. Note that the membership functions are arranged such that they are equally spaced, symmetrical, and that the sum of the membership functions at any one

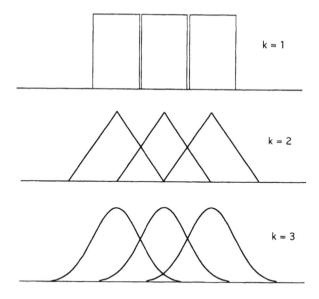

Figure 8.12. B-splines of order 1, 2 and 3.

point along the axis is 1. For the rest of this section the triangular membership functions will be used, so $k = 2$.

The rules and confidence values can be expressed as follows:

$$IF \ (x \ is \ X_i) \ THEN \ (y \ is \ Y_j)(c_{i,j})$$

where x is the input variable, X_i is a fuzzy term (with a total of p terms), y is the output variable, Y_j is a fuzzy term (with a total of q terms), and c_{ij} is the confidence in that rule. This is similar to the TSK method so far, with the exception that a confidence value is included. One other difference is that logical combinations in rules, such as AND and OR, are treated algebraically by taking the product and sum respectively, rather than the MIN and MAX operators.

The goal is therefore to produce rules in this form. The way this is done is to find a set of weights governed by

$$y(x) = \sum_{i=1}^{p} \mu x_i(x) w_i. \tag{8.29}$$

In other words, the desired output values are found as a linear combination of the input membership values. The weights can be found using a variety of methods including the least-means-squared algorithm (LMS), so that the problem could be expressed in terms of conventional neural networks such as the perceptron with a linear output.

Having found the set of weights, the next step is to convert the weights into rules and confidence values. Each weight is regarded as an output value and applied to the output membership functions. The membership values produced are the confidence values for each rule. So, for example, if weight w_i is applied to the output membership functions, there will be q membership values produced. These are $w_i : c_{i,1}, c_{i,2}, \ldots, c_{i,q}$. Each of these terms corresponds to a rule and is the confidence value for that rule. So, for example, $c_{1,2}$ corresponds to the rule:

$$IF \ (x \ is \ X_1) \ THEN \ (y \ is \ Y_2)(c_{1,2}).$$

Since there are p input membership terms, the total number of rules and confidence values produced is $p \times q$. However, the membership functions are defined in such a way that at any point along the axis, only k functions overlap. In this example where triangular functions are used, at most two membership functions overlap, so for any value of w_i there will be at most two non-zero confidence values, and therefore two rules.

The confidence values could be considered similar to the weights suggested for the rules in the AFAM method, whereas the training is more like the TKS training of linear functions. In any case, this method has the ability to generate new rules and to adapt to a changing environment.

8.7 Summary

This chapter has given a brief description of some of the methods that are currently being tried in the area of intelligent control. The work on neural networks has so far generated interesting solutions, but to some extent is still based mainly in the laboratory. Fuzzy logic, on the other hand, has been applied to a range of practical industrial problems and has been very successful. The merger of the two techniques is still in its infancy, but it does appear to offer a way forward to a standard method of controlling complex systems. The hope is that it will prove to be as successful as linear control theory has been in the past.

Chapter 9

Image Compression using Neural Networks

Christopher Cramer[1] and Erol Gelenbe
Duke University, USA
[1] cec@ee.duke.edu

9.1 Introduction

Data compression is increasingly becoming an important subject in all areas of computing and communications. While it is true that network speeds have significantly increased, and that the price of disk storage has decreased dramatically, new computer applications which include multimedia documents with very large data set requirements are constantly pushing the limits of these evolving boundaries. For this reason, data compression will always be important, regardless of the current state-of-the-art in network and storage technologies.

There are two broad approaches to data compression methods, *lossless* and *lossy* compression. As the name implies, lossless compression methods allow the data to be decompressed without loss. In other words, the data that have been compressed can be recovered exactly through the decompression process. On the other hand, lossy compression results in the decompressed data being similar to, but not exactly the same as the original. Each type of compression has its own uses. Lossless compression is basically the best way to compress text or compiled computer programs; whereas lossy compression (due to its higher compression ratios) is better suited for audio, image and video data

Current research in neural network compression focuses almost exclusively on lossy compression. The primary reason for this is that neural networks provide optimized approximations in the applications where they are used. When they are used for compression and decompression of data, they yield no exact guarantees on the resulting quality. In many applications it is good enough for a method to yield the correct decompressed data a certain percentage of the time; however, imagine trying to read an article in which 5% of the words are incorrect. Clearly, even in textual applications, intelligent correction after decompression may result in highly accurate or at least acceptable renditions of the original text. Thus the range of applications of approximate techniques such as neural networks is also broadening. However, imaging remains an area where lossy compression/decompression techniques are generally well accepted. Thus this chapter will discuss the use of neural networks for the lossy compression and decompression of grayscale images.

The reason that grayscale images are primarily discussed here is that any color image can be separated into multiple component images. Each component can then be compressed as if it were a single grayscale image. Although these resulting grayscale images can be treated as being distinct, the correlation between the component grayscale images of a color image can also be used to achieve a higher level of compression.

The following sections will first discuss the metrics used in image compression, then we will review some stand-alone neural compression techniques, and finally consider the use of neural networks to augment standard compression methods.

9.2   Metrics

There are several ways of expressing the degree of compression achieved by a given compression scheme, and the resulting quality when the data is decompressed. One approach expresses compression by stating the number of bits in the compressed image required to store each pixel in the original image. Since this number does not express how many bits were originally required to store each pixel, compression results expressed in bits per pixel can be confusing. For this reason, results in this chapter will be presented in terms of the *compression ratio* defined as:

$$\text{compression ratio} = \frac{\text{bits in original image}}{\text{bits in compressed image}}. \tag{9.1}$$

The compression ratio gives a more accurate representation of the amount of compression achieved and is essentially transparent to the differences between color and grayscale images.

As was previously explained, lossy compression schemes produce images which upon reconstruction are not identical to the original image. To express the image quality generated by a particular lossy compression method for a given picture, the *peak signal to noise ratio* (PSNR) in decibels is used. In this metric, the original image is viewed as the 'signal', while the decompressed image (which is assumed to be of the same size as the original image) is considered to represent the 'signal plus the noise'. This metric is relatively crude since it does not specifically relate to the resulting decompressed image quality as perceived by a human viewer, or even by a specific technical application which may be the end user. However, it has the advantage of simplicity. It is defined as:

$$\text{PSNR} = 10\log_{10}\left(\frac{xy255^2}{\sum_{i=1}^{x}\sum_{j=1}^{y}(I_c(i,j) - I_o(i,j))^2}\right) \qquad (9.2)$$

where x and y represent the size of the image in the horizontal and vertical dimensions, I_o is the original image and I_c is the reconstructed (decompressed) image. Here 255 is used as the signal level because it is the maximum gray level value that can be obtained by using an 8 bit pixel. This peak value is used as the signal, as opposed to the actual pixel value, in order to keep the PSNR for a given compression technique relatively constant for images of varying brightness.

For example, consider an identical compression/decompression scheme which is used for two different images. Suppose that the first image contains pixels which all have a grayscale value of 5. The second image contains pixels which all have a grayscale value of 255. If the compression scheme gives an average error of five gray levels for each image then the PSNR for both images is 34 dB. However, if the signal to noise ratio (SNR) were used, the first image would have an SNR of 0 dB, and the second image would have an SNR of 34 dB. Therefore, the quality of the compression scheme would appear to be based upon the average gray level of the image even if the error it introduced was constant.

9.3   Basic image compression

The simplest neural network based image compression uses a feed-forward network with the topology seen in figure 9.1.

The compressor and decompressor are a single neural network where the compressed data are found at the output of the hidden neurons and the decompressed data are found at the output of the output neurons. This network has an equal number of input and output neurons and some smaller number of hidden neurons. The number of input (and output) neurons corresponds to the size of the image block that you want to compress—one neuron per pixel. For example, to compress an 8×8 block of an image would require 64 input and output neurons. The number of hidden neurons in the network is determined by the number of input and output neurons as well as the desired compression ratio. The compression ratio in neural compression is the ratio of input to hidden neurons. So, if a network was designed to compress 8×8 blocks and the desired compression ratio is 4:1 then the necessary number of hidden neurons is 16 (64:16 = 4:1).

Once the proper network architecture has been designed, the network must be trained. Input to the network is blocks of a training image which have been scaled from the pixel range (usually 0–255) to the input range of the neural network, often 0–1.0. The network is then trained to produce, at the output

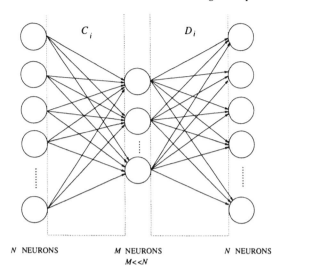

C_i D_i

N NEURONS   M NEURONS   N NEURONS
                   M<<N

Figure 9.1. Simple neural compressor/decompressor.

neurons, what is seen at the input neurons. Once the network has been trained each block of the image is presented to the input neurons. The output of the hidden layer is then taken and used as the compressed data. It should be noted that this output must be quantized to the same number of bits as were in the original pixels in order to obtain the desired compression ratio.

To illustrate this consider the hypothetical network just discussed. Assume that the images being compressed are 8 bit (0 to 255 range). Since the network has 64 inputs the total number of bits before compression is 512 bits per image block. The network has 16 hidden neurons (since it was designed for 4:1 compression). If we take the output of each hidden neuron as a single precision floating point number (32 bits), then the output of the hidden layer requires 512 bits to represent (16×32 bits = 512 bits). This means that the information has not been compressed at all. Therefore, the output of the hidden neurons should be expressed using 8 bit fixed point representation (16×8 bits = 128 bits).

Now that the image has been compressed the natural thing to do is to decompress it. In decompression the compressed data are converted from the 8 bit fixed point representation back to floating point. The outputs of the hidden layer are then set to these data (we now ignore the input layer). The output of the output layer is then computed based on the weights from the hidden to output layers and the new hidden layer outputs. The output of the output layer is, of course, still in the range used by the network (i.e., 0 to 1.0), so the outputs must be scaled back to the 0 to 255 range (assuming 8 bit pixels). Once this operation has been performed on all of the compressed data blocks in the image the image is decompressed and can be displayed.

At this point, several questions are likely to come up. The first being, is it necessary to train a neural network for each image being compressed? Fortunately, it is not necessary. The idea is to get a network that generalizes enough to compress many different types of image block. Since there is only one neural network, the network is considered part of the codec (coder/decoder) so that the compressor and the decompressor are using the same network, i.e., it is not necessary to train and transmit the network weights for each compressed image. Another common question is: suppose a network is trained on one image, will portions of that image show up when compressing other images? For example, if the network was trained with the famous 'Lena' image (figure 9.2), will Lena's face show up when decompressing the 'Peppers' image (figure 9.3)? This does not occur because the network only compresses blocks of the image, and should only map similar blocks to the same compressed data. Therefore, while some blocks of the decompressed 'Peppers' image may resemble blocks of the decompressed 'Lena' image, the total image will look like 'Peppers' and not 'Lena'.

9.3.1 Basic compression implementations

One of the first uses of this basic method was by Sonehara *et al* [92] in which the authors use an N-CUBE massively parallel computer for implementation. In this paper the authors use a 0 to 1 sigmoid and scale their input data within the range [0, 1.0]. The results show that this basic form of neural image compression achieves very poor image quality, not only for the untrained images, but for trained images as well. The authors present their results in terms of the signal to noise ratio as opposed to the peak signal to noise ratio so it is difficult to compare results with other methods. However, in our tests, this basic method achieves the results summarized in table 9.1, where the trained image is 'Lena' and the untrained images are 'Peppers' and 'Girl'. It is interesting to point out that none of these results are as good the result that could be achieved by simply taking the average of each 8×8 block and using that value in all 64 pixels of the decompressed block which gives 64:1 compression at 25 dB. The reason why this should be so will be discussed in a later section.

Table 9.1. Results of the most basic neural compression method.

Compression ratio	Lena	Peppers	Girl
4:1	24.358	23.023	22.141
8:1	24.384	23.045	22.170
16:1	24.384	23.041	22.207
32:1	15.754	14.787	13.173

In addition to implementing the most basic neural compression technique, the authors experiment with two other important ideas in compression. The

Figure 9.2. The LENA image.

Figure 9.3. The PEPPERS image.

first idea is that of quantization of the hidden layer compressed data. As was previously mentioned, the hidden layer outputs should be quantized to some level; if this level uses 8 bits then the compression ratio is the ratio of the number of input neurons to the number of hidden neurons. The authors varied the quantization from 2 bits to 10 bits. They found that while there was a significant performance loss by using only two bits (an increase of 4 in compression ratio) there was no significant performance gain in using more than 6 bits. This implies that an increase in compression ratio by a factor of $\frac{4}{3}$ can be achieved with no significant loss in image quality. A strikingly similar result is seen by Sicuranza *et al* [86] where the quantization is varied between 1 and 8 bits per neuron resulting in little added benefit in using more than 5 bits per neuron.

The second important idea which was tested is the use of different starting weights for the neural network. The authors tried initializing the weights to a subset of the 'weights' used in the discrete cosine transform (DCT), first discussed by Ahmed *et al* [2]. It was found that after training the new weight set bore little resemblance to the DCT weights. Furthermore, an increase in reconstructed image quality was not observed.

9.3.2   CODECs with multiple compression networks

De-Fu Cai and Ming Zhou [13] expanded upon the basic compression method by using two neural networks, the first for high-activity blocks and the second for low-activity blocks. The low-activity network provides a higher compression ratio by using fewer hidden neurons. The high-activity network yields a smaller compression ratio but should better reproduce more complicated image blocks.

The authors begin by first taking image blocks, x, with elements x_i and finding the average of each block, $\bar{x}$. The difference between each block and its average is then found, $dx_i = x_i - \bar{x}$. To determine the activity E of each block, the average of the absolute value of dx is found. Blocks with an E higher than some threshold are compressed by the high-activity network, while other blocks are compressed by the low-activity network. The block dx scaled between -0.5 and 0.5 is compressed, rather than the original block. The sigmoid used for each network is also scaled between -0.5 and 0.5. Once blocks are classified, training proceeds in the normal fashion, and a much improved neural compression technique results. Not only are the PSNR values for the trained images higher, but much of the difference in reconstructed quality between the trained and untrained images disappears. It should be noted that there is an overhead associated with this method. It is necessary to send the block average $\bar{x}$ as well as the block type (high or low activity) with each block of compressed data. However, even with this overhead, this approach offers a marked improvement over the basic method.

9.3.3   Adaptive methods

Several other authors have found that separating blocks into different classes with a different compressor per class improves performance. Zhong Zheng *et al* [117] divided image blocks into different classes based upon the gray level difference V_d of each block, where for a given block x, $V_d = \max(x) - \min(x)$. Neural networks were assigned to compress blocks with different ranges of the value V_d. Furthermore, the different networks were given a differing number of hidden neurons. The results show that this technique yields modest improvements in the PSNR between 1 and 2 dB.

Marsi *et al* [62] suggested two additional approaches. They divided blocks into classes based upon the activity of the block. A_b, defined as the sum of the pixel activity $A_p(x_{m,n})$ for each pixel $x_{m,n}$ in the block, where:

$$A_p(x_{m,n}) = \sum_{i=-1}^{1} \sum_{j=-1}^{1} (x_{m,n} - x_{m-i,n-j}^2) \quad \text{for} \quad (i, j) \neq (0, 0). \tag{9.3}$$

In the first approach by Marsi *et al*, the image blocks are divided into four different categories based upon the level of activity. A neural network is trained to compress each different category, and networks for compressing low-activity blocks achieve greater compression than those used for higher-activity blocks, yielding significant improvement over the standard method. The second approach again uses the activity of a block to classify it into three categories: low, medium, and high activity. High-activity blocks are further subdivided into four categories based upon the orientation of the block: horizontal (h), vertical (v), and the two diagonal directions (d and e), determined by evaluating the following functions. The function with the minimum value determines the block's orientation:

$$B_h = \frac{1}{M(M-1)} \sum_{m=1}^{M} \sum_{n=1}^{M-1} (x_{m,n} - x_{m,n+1})^2 \tag{9.4}$$

$$B_v = \frac{1}{M(M-1)} \sum_{m=1}^{M-1} \sum_{n=1}^{M} (x_{m,n} - x_{m+1,n})^2 \tag{9.5}$$

$$B_d = \frac{1}{(M-1)^2} \sum_{m=1}^{M-1} \sum_{n=1}^{M-1} (x_{m,n} - x_{m+1,n+1})^2 \tag{9.6}$$

$$B_e = \frac{1}{(M-1)^2} \sum_{m=2}^{M} \sum_{n=1}^{M-1} (x_{m,n} - x_{m-1,n+1})^2. \tag{9.7}$$

There are now six block categories and six corresponding neural networks. The second adaptive approach gives a better visual quality at a given compression ratio but did not yield increased PSNR, demonstrating the inadequacy of PSNR as a measure of visual quality.

Qiu *et al* [77] expand upon the work by Marsi *et al* [62] by classifying each block solely by its orientation: two horizontal, two vertical, four diagonal, and one solid orientations are considered (see figure 9.4). Nine different neural networks were used in order to compress the various orientations a block could have.

Figure 9.4. Possible block orientations.

To determine the orientation of a block b of size $m \times n$, the authors used two directional derivative approximations G_x and G_y:

$$G_x = \frac{2}{m \times n} \left[\sum_{i=m/2}^{m-1} \sum_{j=0}^{n-1} b_{ij} - \sum_{i=0}^{(m/2)-1} \sum_{j=0}^{n-1} b_{ij} \right] \tag{9.8}$$

$$G_y = \frac{2}{m \times n} \left[\sum_{i=m=0}^{m-1} \sum_{j=n/2}^{n-1} b_{ij} - \sum_{i=0}^{m-1} \sum_{j=0}^{(n/2)-1} b_{ij} \right]. \tag{9.9}$$

The gradient magnitude and gradient orientations are defined as

$$|G| = \sqrt{(G_x^2 + G_y^2)} \quad \text{and} \quad \angle G = \tan^{-1} \left[\frac{G_y}{G_x} \right]$$

respectively. To determine whether a block is a shaded block the gradient magnitude is compared to some threshold, and those with a magnitude less than the threshold are considered shaded blocks. Directional blocks are then categorized based upon the gradient orientation. Once the blocks have been classified, the authors determine the mean gray level of the block and compute a new block which is the original block minus the block mean. This new block is then passed to the appropriate neural compressor. Therefore, the compressed data in this method is the block mean, the orientation number of the block and the output of the hidden layer.

This method yielded some of the best results, even though to accommodate the extra overhead the authors used two hidden neurons fewer than in the methods they were comparing with. The results show an improvement of approximately 1 dB over their baseline technique which is an improved form of the approach used by Sonehara *et al* [92].

9.3.4 Different neural network models

While all of the other papers discussed in this section have used the standard back-propagation model, other neural network models have also been used. Setiono and Guojun Lu [84] use a neural network that has the ability to add hidden nodes if the quality currently being achieved is not good enough. Training in this network begins by creating a basic neural network compressor/decompressor with only one hidden neuron ($h = 1$). The network is then trained until the reproduced image quality has been maximized. If this level of quality meets some given threshold, then training is stopped. If not, a new hidden neuron is added. The initial values of weights from the input layer to the new hidden neuron are chosen at random and denoted by w^{h+1}. In order to insure that adding a new neuron will not increase the error associated with the neural network, the following function is used to define the weights from the hidden layer to the output layer:

$$v^{h+1} = \frac{-\sum_{i=1}^{k} \sigma((x^i)^T w^{h+1}) e^{-1}}{\sum_{i=1}^{k} (\sigma((x^i)^T w^{h+1}))^2} \tag{9.10}$$

where x^i is an input vector to the neural network, k is the number of input vectors being trained with, and $\sigma(z)$ is the neural network sigmoid function. Once the new neuron has been added ($h \leftarrow h+1$), and its weights determined the training of the network continues. Another interesting feature of this approach is the training method used. Whereas most neural networks use the back-propagation training algorithm, the neural networks used by here were trained using a quasi-Newton method in order to speed up the training time. Results show definite improvement over the traditional neural compression technique. Furthermore, this method is able to specify a desired quality level that the neural network attempts to achieve automatically.

Kohno *et al* [51] use a neural network which has the ability to learn its own sigmoid function defined as:

$$f(x) = \begin{cases} 2/[1 + \exp(-ax)] - 1 & \text{if } |x| \geq c \\ \\ (2/[1 + \exp(-a)] - 1)\,x/c & \text{if } |x| < c \end{cases}$$

where the parameter c defines the linear region of the sigmoid ($c = 1$ in this chapter) and a determines the non-linearity. The weight parameter for the jth neuron in the kth layer $a(k, j)$ can then be learned using the following update rule:

$$a(k, j) \leftarrow a(k, j) - \eta \frac{\partial E}{\partial a(k, j)} \tag{9.11}$$

$$\frac{\partial E}{\partial a(k, j)} = \begin{cases} \delta(k, j) I(k, j)/a(k, j) & \text{if } |x| \geq c \\ \\ \delta(k, j)[I(k, j)^2 - O(k, j)^2]/2i(k, j) & \text{if } |x| < c \end{cases} \tag{9.12}$$

where $I(k, j)$ is the summed input to the jth neuron in the kth layer, $\delta(k, j) = \partial E/\partial I(k, j)$, and $O(k, j)$ is the output of neuron (k, j). The primary use of the constant $a(k, j)$ is to shorten the learning time. This neural network and its learning rule do not perform significantly better than most other basic compression methods. However, other adaptive methods using the non-linear factor which do increase performance are also discussed. In one approach, a network with a fixed adaptive parameter ($a = 1$) for all neurons is used. Once the weights have been trained, the authors pick a small set of possible values of a for all neurons in the hidden and output layers. In this case the values were: 0.9, 1.0, 1.1 and 1.2. To compress a block of an image, the value of a for the hidden layer and the value of a for the output layer were chosen to give the best performance. The index values for the two parameters were then sent to the decompressor along with the compressed data. The performance yielded by this adaptive method is only slightly better than a standard compressor with results generally improved by approximately 0.5 dB for a given compression ratio. In a second adaptive approach, a set of weights is learned for the neural network with $a = 1$ for all neurons. Once the weights are learned, for each block to be compressed a is learned for all neurons in the hidden layer. The values of these parameters are then quantized and transmitted along with the compressed data to the decompressor. This method results in a substantially longer compression time since compressing each block requires some training, while the results are only fractionally better than the previous adaptive method.

9.4   The random neural network model

The previous discussion on neural image compression is based upon a standard feed-forward connectionist network with some slight variations. On the other hand, Gelenbe *et al* [38, 39] and Cramer *et al* [20] discuss compression based upon an entirely different neural network model known as the *random neural network* (RNN). The compression method used is essentially the same as used by Sonehara *et al* [92], the inputs are scaled between 0 and 1, and the network is trained to produce at the output neurons what it sees at the inputs. The results, however, are substantially different. The PSNR for this method is anywhere from 3 to 12 dB higher than the baseline. This represents as much as a 75% reduction in error. The difference is, of course, the neural network involved.

In the random neural network model developed by Gelenbe [34, 35, 37, 36] signals in the form of spikes of unit amplitude circulate among the neurons. Positive signals represent excitation and negative signals represent inhibition. Each neuron's state is a non-negative integer called its potential, which increases when an excitation signal arrives at it, and decreases when an inhibition signal arrives.

The state of the n-neuron network at time t, is represented by the vector of non-negative integers $k(t) = (k_1(t), \ldots, k_n(t))$, where $k_i(t)$ is the potential or internal integer state of neuron i. We will denote by k and k_i, respectively, arbitrary values of the state vector and of the ith neuron's state.

Neuron i will 'fire' (i.e., become excited and send out spikes) if its potential is *positive*. The spikes will then be sent out at a rate $r(i)$, with independent, identically and exponentially distributed inter-spike intervals. Thus $r(i)$ is the rate or parameter of the exponential distribution of the times between successive spike emissions when neuron i is 'excited'. The spikes being sent out by neuron i are sent as excitatory or inhibitory spikes to other neurons, or they can be lost as in the case of a neuron whose potential is depleted.

An excitatory spike is interpreted as a '+1' signal at a receiving neuron, while an inhibitory spike is interpreted as a '−1' signal. Neural potential also decreases when the neuron fires. Thus a neuron i emitting a spike, whether it be an excitation or an inhibition, will lose potential of one unit, going from state value k_i to the state of value $k_i - 1$.

Spikes will go from neuron i to some neuron j with probability $p^+(i, j)$ as excitatory signals, or with probability $p^-(i, j)$ as inhibitory signals. A neuron may also send signals out of the network with probability $d(i)$, and all these probabilities will obey the constraint:

$$d(i) + \sum_{j=1}^{n} [p^+(i, j) + p^-(i, j)] = 1. \tag{9.13}$$

Let us denote by

$$w_{ij}^+ = r(i) \, p^+(i, j) \tag{9.14}$$

and

$$w_{ij}^- = r(i) \, p^-(i, j). \tag{9.15}$$

Here the w play a role similar to that of the synaptic weights in connectionist models, though they specifically represent *rates* of excitatory and inhibitory spike emission. They are thus *non-negative* quantities. Exogenous (i.e., those coming from the 'outside world') excitatory and inhibitory signals also can arrive at neuron i according to Poisson (random) processes of rates $\Lambda(i)$, $\lambda(i)$, respectively. In general, this is a 'recurrent network' model, i.e., one in which feedback loops of arbitrary topology are allowed.

Computations related to this model are based on the probability distribution of network state $p(k, t) = \Pr[k(t) = k]$, and on the marginal probability that neuron i is excited $q_i(t) = \Pr[k_i(t) > 0]$. As a consequence of the probabilistic assumptions detailed above, the time-dependent behavior of the model is described by an infinite system of *Chapman–Kolmogorov* equations for discrete state-space continuous time Markovian systems. In this review we do not detail these state transition equations and refer the interested reader to the work by Gelenbe [34, 36].

Information in this model is carried by the *frequencies* $\{w_{ij}^+, w_{ij}^-\}$ at which spikes travel. In turn, each neuron behaves as a non-linear *frequency demodulator* since it transforms the incoming excitatory and inhibitory spike trains' rates into an 'amplitude', which is $q_i(t)$ the probability that neuron i is excited at time t. Each neuron of this model is also a frequency modulator since neuron i sends out excitatory and inhibitory spikes at rates (or frequencies) $q_i(t)r(i)p^+(i, j)$ and $q_i(t)r(i)p^-(i, j)$ to any neuron j.

The stationary probability distribution associated with the model will be the quantity we use throughout the computations:

$$p(k) = \lim_{t \to \infty} p(k, t) \qquad q_i = \lim_{t \to \infty} q_i(t) \qquad i = 1, \ldots, n. \tag{9.16}$$

It is obtained via the following result derived by Gelenbe [34, 36].

The stationary probability that any neuron i is excited (denoted by q_i) is given by

$$q_i = \lambda^+(i)/[r(i) + \lambda^-(i)] \tag{9.17}$$

where the $\lambda^+(i), \lambda^-(i)$ for $i = 1, \ldots, n$ satisfy the system of nonlinear simultaneous equations:

$$\lambda^+(i) = \sum_j q_j r(j) p^+(j, i) + \Lambda(i) \qquad \lambda^-(i) = \sum_j q_j r(j) p^-(j, i) + \lambda(i).$$

(9.18)

Furthermore if the resulting $q_i < 1$, then

$$p(k) = \prod_{i=1}^{n} [1 - q_i] q_i^{k_i}.$$

(9.19)

Throughout the application to image compression, we will only make use of equations (9.17) and (9.18), and of the RNN's learning algorithm proposed by Gelenbe [36], which is not discussed in this survey.

9.4.1 Image compression with the RNN

In using the RNN for image compression, the connectivity of the network is simplified so that the architecture is that of figure 9.1. The Λ_i values for the input neurons are set to the scaled blocks of the image. The 'outputs' of the network are the probabilities that the output neurons are excited. The results of this compression are given in table 9.2 for trained (Lena) and untrained (Peppers and Girl) images. Clearly, the RNN easily provides an improvement of well over 10 dB in resulting decompressed image quality as compared to the basic connectionist network results summarized in table 9.2.

Table 9.2. Image quality for varying levels of compression on the trained and untrained images, using the RNN method.

Compression ratio	Lena	Peppers	Girl
4:1	33.606	32.162	33.240
8:1	31.002	29.231	29.685
16:1	29.301	28.236	28.309
32:1	27.266	25.995	26.141

In view of these results, a discussion of the reasons why the RNN provides this performance improvement is in order. De-Fu Cai and Ming Zhou [13] and Qiu *et al* [77] have explained the success of their compression schemes by suggesting that the techniques they use help provide fewer conflicting blocks to the neural networks. The most common of these techniques is to normalize the block by its mean. This suggests that input scaling is a major factor in the success of neural compression schemes which use a positive/negative scale by subtracting the block mean.

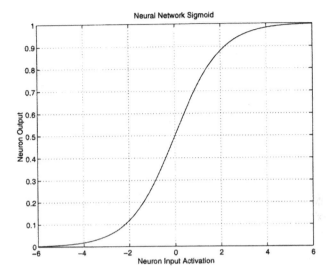

Figure 9.5. Typical neural network sigmoid function.

Let us attempt an intuitive explanation to this apparent success. Consider a given block B_1 which is to be compressed by a single neural network using $[0, 1]$ input scaling. In this block there may be pixels with gray levels less than 127 and some with gray levels greater than 127. In order for an output neuron to produce an output value less than 0.5, corresponding to a grayscale value less than 127, the input to it must be negative (see figure 9.5). Since all of the inputs to the neural network are positive, some of the weights must be negative in order for the network to reproduce pixels with values less than 127, because the neurons in the network require a negative summed input in order to produce an output value less than 0.5. After the neuron has been trained with some block B_1, block B_2 is presented to the network. Suppose that his block has pixel values of $B_2(i, j) = 255 - B_1(i, j)$, i.e., all positions in block B_1 with pixel values less than 127 have values greater than 127 in block B_2. When trying to learn this block, the network will find that all of the negative weights designed to produce a pixel value less than 127 in B_1 now must be positive in order to produce pixel values greater than 127 in block B_2. Obviously, these two blocks cannot both be learned well, which is why the best one may achieve with this input representation is to have the network produce an output which is homogeneous across any given block and is essentially equal to the block's mean. This explains why the standard method with 0 to 1 scaling produces PSNR's less than those achievable by assigning to the pixels in a given block the value of the block's mean.

In contrast, input scaling within the range $[-1.0, 1.0]$ allows a neuron to have all positive weights and still receive a negative input sum. This allows the neuron to produce a grayscale value less than 127. However, if the next block requires the same output neuron to have a value greater than 127, the fact that all weights are positive means that the input to the output neuron can be positive without conflicting with the previous weight assignment. Therefore, presenting different blocks to the network allows it to fine tune its reconstructed blocks without completely erasing the previous training.

This also helps explain why the random neural network performs as well as it does using only excitatory inputs. The 'sigmoid' of the RNN with no external inhibition external input (which would appear as a positive term λ_i in the denominator of q_i) is:

$$q_i = \frac{\Lambda_i + \sum_{j=1}^{n} q_j w_{ji}^{+}}{r_i + \sum_{j=1}^{n} q_j w_{ji}^{-}} \qquad (9.20)$$

where Λ_i is the excitatory input, and w_{ji}^{+} and w_{ji}^{-} are the excitation and inhibition weights from neuron j to neuron i. Note again that all weights are positive. Since this network only uses positive valued signals, the problem of producing a neuron output value less than 0.5 by obtaining a summed input value less than 0 does not exist. Thus scaling the input to $[0, 1]$ in this network has the same generality as scaling the input to $[-1, 1]$ in a connectionist network.

9.5 Vector quantization

The other major use of neural networks in image compression has been as a complement to the well known technique known as vector quantization. In vector quantization, the first step is to design what is known as a 'codebook'. The codebook is a list of blocks that can be used to *decompress* a given image. The goal of compression, therefore, is to find the best match in the codebook to a given block of the image. The compressed data are the index to the best match in the codebook. Compression is achieved because the number of blocks in the codebook can be indexed by many fewer bits than there are in an image block. For example, if a 256-block codebook is designed, then each index to the codebook is only eight bits long. If this codebook is used upon 4×4 blocks of an eight-bit grayscale image then the compression ratio is 16:1.

The main difficulty with this compression method is designing the codebook itself, which is an optimization problem with an almost infinite solution space. To overcome this problem, Nasrabadi and Yushu Feng [67] use a Kohonen network (figure 9.6) to design the codebook. This network will have the same number of input neurons as the number of pixels in a block (K). The number of clusters, or output neurons, in the Kohonen network is set to N the number of blocks in the codebook.

To train the neural network, one first initializes all of the weights between the input and output layers to some small random value. An input block from the training image is chosen. Denote the vectorization of this block x with

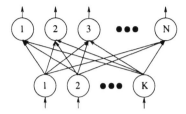

Figure 9.6. Kohonen network for vector quantization.

elements x_i. For each output neuron j, $j = 0, ..., N$, the distance between the neuron and the current vector is computed:

$$d_j = \sqrt{\sum_{i=1}^{K}(x_i - w_{ij}^2)} \qquad (9.21)$$

and the neuron j^* with the minimum distance d_j is selected to be the 'winner', and its weights adapted according to the rule:

$$w_{ij} \leftarrow w_{ij} - \alpha(x_i - w_{ij}) \qquad (9.22)$$

where $\alpha > 0$ is a learning constant. Once the winning neuron has been modified, a new input vector is selected and the process is repeated.

Nasrabadi and Yushu Feng designed two codebooks; the first is for high-variance blocks and uses 284 entries. The second is for low-variance blocks and only contains 64 entries. The compression results of these codebooks are compared to those of codebooks designed by a standard method (LBG) developed by Linde *et al* [59]. It was found that these novel codebooks yielded a quality increase of approximately 1 dB over the standard codebooks for the same level of compression.

9.5.1 Adaptive vector quantization

Cheng-Chang Lu and Yong Ho Shin [61] use a standard Kohonen network to design codebooks for image compression. What makes this method different from the method presented by Nasrabadi and Yushu Feng [67] is that the authors use four different codebooks. The first codebook is designed for low-activity blocks. Blocks of higher activity are further subdivided into horizontal, vertical and diagonal categories, corresponding to the primary direction of the block.

To determine the direction of a block, the authors use a feed-forward multi-layer perceptron network. This network has been trained using a pre-defined set of binary edge blocks, as opposed to real image blocks for which the orientation has been determined mathematically.

Results from this method are compared with the LBG algorithm and are found to exceed the LBG codebook by approximately 0.5 dB for a given compression ratio. This could perhaps be improved upon if the authors were to have taken an approach more similar to that taken by Qiu *et al* [77].

9.5.2 Adaptive Kohonen networks

One of the major problems with the Kohonen network, in general, is that some neurons are under-utilized. In order to handle this problem for generating vector quantization codebooks, Hsien-Chung Wei *et al* [101] suggest an adaptation of the Kohonen network called the hierarchical Kohonen neural network (HKNN). In this method, the authors begin by designing a small codebook using the standard Kohonen network. Once this codebook has been designed, the authors search the training set to find the block clusters corresponding to each index in the codebook. The 16 indices having the highest cluster error (mean error over all training blocks matching the given index) are divided into four new clusters and training continues. This process is repeated three more times, so that if the initial codebook contains 64 vectors, the final codebook will contain 256 vectors. The authors found that this method yielded a substantial increase in the reconstructed image quality as compared to both the LBG algorithm and a standard Kohonen designed codebook.

Oscal Chen *et al* [15] took a similar approach to codebook design. The first primary difference is that the authors considered the frequency of a neuron's usage as well as the distance to the current input vector to decide which neuron's weights would be updated. The new distance function to minimize is:

$$FD_i = \left(1 + \frac{F_i}{F_{thd}}\right) \sqrt{\sum_{j=1}^{k}(X_{pj} - W_{ij})^2} \qquad (9.23)$$

where X_p is the current input vector, F_i is the usage frequency of neuron i, F_{thd} is a frequency threshold and W_{ij} is the weight vector for neuron i.

The introduction of the usage frequency to the distance equation prevents neurons from being completely unused or vastly over-used. If a codebook is designed using only this equation, the authors' call it a 1-path codebook. N-path codebooks begin as 1-path codebooks; however, once the codebook is designed under-used code vectors are eliminated and over-used code vectors are split into multiple code vectors and training continues. This process is repeated $N - 1$ times in an N-path codebook. Results indicate that using a 4-path or 5-path codebook can yield an increase in quality of approximately 1.5 dB over using a 1-path, or standard, Kohonen codebook.

9.5.3 VLSI Kohonen networks

The other major problem with vector quantization is that once the codebook has been designed, for each block of an image, the entire codebook must be searched in order to find the best match. This is another area in which a

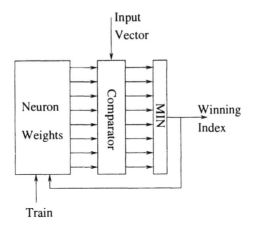

Figure 9.7. Simple block diagram of a hardware vector quantization scheme.

Kohonen network is a great improvement over traditional vector quantization methods. When implemented in hardware (figure 9.7) there is little difference between the neural network for learning and for compression. The same winner-takes-all approach is utilized. However, in training the winning neuron (or code vector) is updated; whereas in compression, the winning code vector index is stored or transmitted as compressed data. This approach has been used by Bing Sheu and Wai-Chi Fang [85] where the authors implemented the basic Kohonen learning rule (equation (9.21)), and in Wai-Chi Fang *et al* [31, 32] where the authors implemented the usage frequency learning rule (equation (9.23)).

9.5.4 Predictive vector quantization

Another approach to the vector quantization compression method is the idea of predictive vector quantization (PVQ). In PVQ, the current image block is first predicted in some fashion based upon the already encoded (or decoded) blocks (figure 9.8). The difference between the predicted and actual blocks is then compressed by means of vector quantization. It has been found that this method allows for a smaller codebook to be used in order to achieve the same quality of reconstructed image.

The main problem with PVQ is the design of the predictor. The easiest and most widely used predictor is the simple linear predictor. However, the linear predictor exhibits poor performance in predicting edge blocks. In order to achieve better prediction, non-linear predictors must be used. Unfortunately, non-linear predictors are difficult to design because of the high complexity involved. Rizvi *et al* [78] examined the use of several neural networks, the multi-layered perceptron network, the functional link network and the radial-basis function network, as predictors.

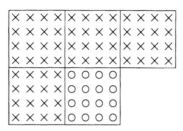

$\times$ Pixels in previous blocks
$\circ$ Pixels in current block

Figure 9.8. Relative position of the current block to the predicting blocks.

Since the blocks in this chapter are of size 4×4, there are 16 pixels that must be predicted. This determines the size of the input and output layers of the neural networks, which must have 64 ($16 \times 4 = 64$) input neurons and 16 output neurons. The networks were then trained to take the four preceding blocks and produce the next block. Once the networks were trained, the resulting predictive images were compared with the linear predictor. It was found that the multi-layered perceptron network performed best yielding a 0.5 dB improvement in prediction over the basic linear predictor.

9.6   Conclusions

The preceding sections have shown various ways in which neural networks can be used for compression, including stand-alone techniques which began simply as seen in Sonehara *et al*, but have evolved into more complicated and effective compression techniques. This has led to one of the more complicated methods in which nine different networks are used for a single level of compression (Qiu *et al*), which yields some of the better reconstructed image qualities. Other methods have been discussed which generate vastly improved images over those seen in Sonehara *et al* by using what is essentially the same method (Gelenbe *et al* and Cramer *et al*).

It has also been shown that neural networks can be used to support and improve upon well known compression techniques. Nasrabadi and Yushu Feng first used a Kohonen network to generate codebooks for vector quantization. Their method proved to be better at generating codebooks than anything else then available. Hsien-Chung Wei *et al* and Oscal Chen *et al* improved upon the standard Kohonen codebook design by using adaptive forms of the Kohonen network.

That is not to say that the techniques presented above are the only areas of research in this field. For example, both direct neural compression and neural assisted vector quantization have been applied to the one dimension data used in speech compression. Daugman [22, 23] has shown that it is possible to represent the Gabor transformation within the context of a relaxation neural network. This method quickly and easily finds Gabor coefficients which can then be entropy encoded and compressed, giving excellent results.

Neural networks have also been used in the compression of digital video sequences. Gelenbe *et al* [39] discuss a basic video compression scheme based on neural networks. This method is capable of achieving the same video quality at a lower bit rate than the current standards (H.261). Cramer *et al* [20] expand upon this basic video compression technique, allowing for substantial increases in the compression ratio with little loss in video quality. To increase the compression ratio, the authors use a temporal sub-sampling and interpolation method which can be performed by neural networks. Other authors, such as Skrzypkowiak and Jain [89, 90], have found ways of using neural networks to quickly compute the motion vectors necessary for MPEG encoding. These examples are but some of the few techniques that exist using neural networks for compression, and will perhaps give the reader an idea of where research in this field is headed.

Acknowledgment

The authors' work reported in this chapter was partially supported by the US Army Research Office under grants Nos DAAH04-96-1-0388 and DAAH04-96-1-0448.

Chapter 10

Oil Spill Detection: a Case Study using Recurrent Artificial Neural Networks

Tom Ziemke[1], Mikael Bodén and Lars Niklasson
University of Skövde, Sweden
[1] tom.ziemke@ida.his.se

10.1 Introduction

Artificial neural networks (ANNs) have been applied to many practical problems. When studying these applications in some detail it becomes apparent that typically a lot of background knowledge is needed, both of the actual task and of ANNs, in order to ensure, or at least allow, a successful application. Typically the available data have to be analyzed, the relevant data have to be extracted and pre-processed, a number of applicable ANN architectures has to be selected/designed, and experiments have to be carried out to evaluate the performance of the different types of ANN in order to find out which architecture is best suited for a particular application. Moreover this design process is often iterative during which the model freedom and learning capabilities of ANNs combined with proper analysis techniques yield further understanding of the problem domain, applicable network architectures and their learning capacities.

The ANN architecture used most (by far) in practical applications is the standard feedforward network trained with the backpropagation algorithm. Due to the large number of applications and the even larger body of theoretical work on this architecture its capacities and limitations are by now rather well understood. Recurrent networks on the other hand, i.e., networks using internal feedback of activation, although no longer uncommon also in practical applications, are still far less well understood, and their use is not seldomly considered to be something of a 'black art'.

Hence, studies of ANNs for the classification of radar measurements, as in many other domains, have mostly been based on feedforward architectures (e.g. [165, 177, 182, 194]), and only relatively few (e.g. [118]) have made use of recurrent architectures. This chapter presents a case study of recurrent ANNs (in comparison to feedforward nets) for oil spill detection from Doppler radar measurements. Although the particular problem at hand is quite specific, it is possible to point to some general, domain-independent issues concerning design and analysis of ANNs. After presenting the problem, we will discuss the choice of data and the need for pre-processing them in order to enhance generalization capabilities. We will then debate the choice of architecture, and finally present and analyze experimental results, showing that the design decisions that, in this case, result in a high degree of robustness to varying radar illumination conditions.

10.2 Problem description

10.2.1 Overview

The overall goal of the work presented here has been the development of reliable and robust ANN models for the detection of oil spills in radar imagery of sea environments. The images are obtained from the measurements of a so-called side looking airborne radar (SLAR), a conventional Doppler radar illuminating sea environments from a plane at a certain distance (see figure 10.1). This chapter presents an extension and further analysis of work presented in [110].

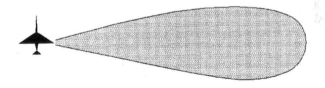

Figure 10.1. Side looking airborne radar (SLAR).

The discrimination between oil and water is possible, at least theoretically, due to the fact that oil dampens the capillary waves present on a sea surface. This has the effect that the surface becomes smoother and acts more like a mirror, i.e., more energy will be reflected away from the radar by an oil covered surface than by pure water (see figure 10.2).

In practice, however, this task, and in particular the detection of small oil spills, is further complicated by the presence of noise and other problems inherent in the nature of radar measurements.

10.2.2 The SLAR simulation model

In this study a simulation model of an SLAR has been used, which is an extension of a model initially developed by Fälldin [144]. The SLAR is a *Doppler radar*,

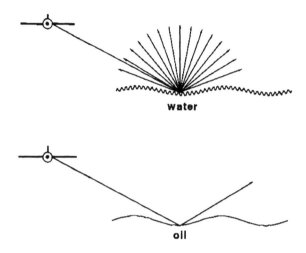

Figure 10.2. Backscattered energy from water and oil surfaces.

i.e., it detects objects by virtue of their velocity with respect to the radar itself. Doppler filtering makes use of the progressive phase shift (*Doppler shift*) of an object's return signal on successively transmitted pulses, which is caused by the object's motion towards or away from the radar (the *radial velocity*).

The SLAR's technical specification (flight speed, frequency of pulses sent out, etc) is such that a rather low resolution of cells of 20 m (in flight direction) × 75 m is achieved. The radar echo from a target area consists of the returns of a multitude of point scatterers (with possibly different radial velocities, e.g. in the case at hand waves moving at slightly different speeds), forming a spectrum of individual phase shifts, the so-called *Doppler spectrum* or *Doppler signature* (see figure 10.3). The model simulates the measurements of 64 Doppler channels (measuring radial velocities in the range between -16 m s^{-1} and $+15.5$ m s^{-1}) for each resolution cell. The above differences in backscattering behavior between oil and water (see figure 10.2) are reflected in their Doppler spectra, as illustrated in figure 10.3 which shows typical signatures for a common sea state.

The mean radial velocity, depending on the waves' translational motion, is of no interest for the classification problem at hand, since it can be assumed to be the same for oil and water. Therefore it would also not make sense to use all 64 Doppler channels as input to a classification network, since it is not the absolute position (or height) of the peaks which is relevant for the classification, but their relation to each other.

More interesting here are the spectral shape and width which to some degree reflect the 'internal dynamics' of an illuminated area/object (see [182]), such that in this case two relevant values/features can be extracted from the Doppler signatures and will be used as input values for the classification:

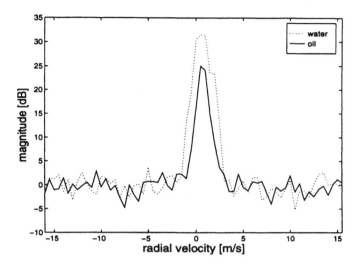

Figure 10.3. Typical Doppler spectra of oil and water.

- Intensity, reflecting the magnitude of the back scattered energy, which, as explained and illustrated above, is lower for oil than for water.
- Spectrum width, i.e., the velocity spread of the individual point scatterers, which, due to the waves' lesser internal motion, is also lower for oil than for water.

The above Doppler signatures for water and oil spills are of course only characteristic for one illumination distance (the work reported here focuses on radar measurements from a distance of 20 km) and one sea state only, i.e., particular wind and weather conditions, resulting in different wave heights, etc. Since we here also want to evaluate the ANNs' robustness to varying wind conditions (i.e., their generalization capability), data from the six different sea states are shown in table 10.1. Note that RCS (radar cross section) is proportional to intensity. The spectrum width values of oil are assumed to be 50% of those of water as stated in table 10.1. (Experiments with values of 30% and 70% were equally successful, showing that this assumption is not critical.)

10.2.3 Data analysis

The difficulty of discriminating oil from water in Doppler radar images is probably best illustrated by an example. Figure 10.4 shows environment A with a size of 1500 m × 1500 m, corresponding to 75 × 20 cells in the radar's resolution. The environment contains two large and five small oil spills (dark areas) surrounded by water (white areas).

The measurements of intensity and spectrum width within this environment in sea state 3 (which is the most common sea state) are shown in figure 10.5.

Table 10.1. Sea states.

Sea state	Wind vel. (m s^{-1})	Radial vel. (m s^{-1})	Wave height (m)	Spectrum width water (m)	RCS water (dB)
1	1–3	0.2	0.1–0.3	0.3	−45
2	3–6	0.7	0.3–0.9	0.6	−41
3	6–8	1.0	0.9–1.5	0.9	−38
4	8–10	1.4	1.5–2.4	1.2	−35
5	10–13	1.7	2.4–3.6	1.6	−33
6	13–15	2.1	3.6–6.0	1.9	−28

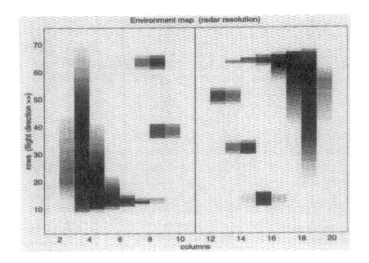

Figure 10.4. Environment A.

Both maps show peaks of low values (dark areas) where the large oil spills have their greatest widths (in flight direction), but the small oil spills cannot be reliably discriminated from the noise.

Apart from the fact that radar measurements are noisy by nature, the task is further complicated by the fact that the radar beam cannot be focused on a single resolution cell. That means that at every time step the radar echo does not only contain the returns of a single resolution cell, but those from a much larger area. The practically relevant area should be approximately that within the 6 dB beam width of 88.5 m, the SLAR model , however, simulates measurements up to the 24 dB beam width of 349 m. This means that the measurements of a particular cell are always influenced by the returns of a number of neighboring cells, as illustrated in figure 10.6. Therefore the radar echo of a single resolution cell containing oil (itself only being 20 m × 75 m in size) will actually only be a

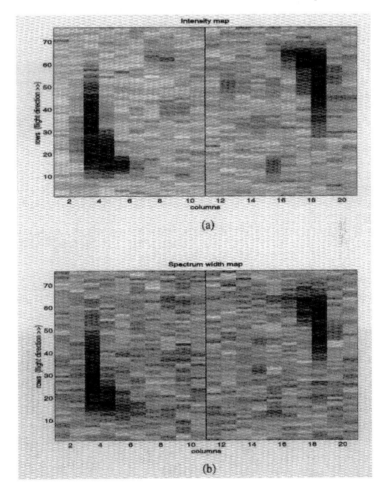

Figure 10.5. Measurements of (a) intensity and (b) spectrum width in environment A (sea state 3).

'pure oil' echo if the whole area covered by the relevant radar beam contains oil (see figure 10.6). This has the effect that at the border of oil spills the echoes of oil and water can get mixed up, and small oil spills (only consisting of 'borders') 'get lost' in their water neighborhood (as the small oil spills in figure 10.5).

Another problem with this data is that, as mentioned above, intensities and spectral widths vary with the sea state (see table 10.1). The general relation between oil and pure water, however, is fairly constant. Therefore, in order to allow learning to generalize across different states, network inputs values are normalized with respect to sea state specific reference values, obtained from an area known to be water. Here the 100 cells in rows 1–5 were used as the water reference area. Hence, input values are computed as relative deviations

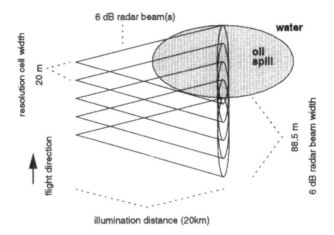

Figure 10.6. Beam width versus cell size.

from these reference values as follows (note that dB is a logarithmic measure, therefore the normalization of intensity values looks a bit more complicated):

$$\Delta i = 10^{(i^M - i^R)/10} - 1 \qquad (10.1)$$

$$\Delta sw = \frac{sw^M - sw^R}{sw^R}. \qquad (10.2)$$

Here Δi is the relative intensity deviation, i^M the measured intensity, and i^R is the reference intensity. Δsw is the relative spectrum width deviation, sw^M the measured spectrum width, and sw^R is the reference spectrum width.

10.2.4 Design of network architecture

So far we have identified the available input data and discussed how to pre-process it. Now we will have to do some initial design decisions for the actual network architecture. If we plot the available data for each resolution cell, i.e., the relative deviations in intensity and spectrum width, together with the cell's (correct) binary classification (using a threshold of 0.5, i.e., 50% oil in a resolution cell), we can identify some clusters in the data (see figure 10.7).

The water cells (marked with + in figure 10.7) cluster around the (water) reference values, i.e., around point $(0, 0)$. The oil cells (marked with o in the figure) mostly have both lower intensity and spectrum width values. But there is a large 'overlap area', that contains both kinds of cell, such that oil and water vectors are not linearly separable (which naturally has an effect on the choice of network architecture, since backpropagation networks without hidden layers cannot solve non-linear problems). The overlap area mainly contains the

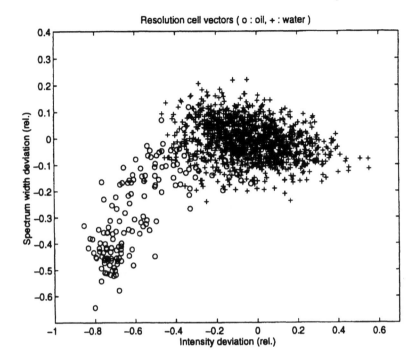

Figure 10.7. Normalized resolution cell vectors for environment A and their binary classification.

vectors of the small oil spill cells. This means that a reliable classification of an individual cell only on the basis of its own values is not possible, due to the noise in the data. Therefore it makes sense to include information from neighboring cells, since resolution cells of a particular type (i.e., water or oil) are likely to have neighboring cells of the same type. Prior analysis [109] showed that taking into account more than the four direct neighbors does not result in better classification performance, and that, even when using the ten-dimensional input (i.e., intensities and spectrum widths for one cell to be classified and its four direct neighbors) the classification problem still remains linearly non-separable, such that we need to use multi-layered networks, on these extended data also.

The *'stretching effect'* in flight direction (i.e., the overlap of radar echoes due to the radar beam being much wider than a single resolution cell, see figure 10.6), does not only pose a problem but could also be used to identify the small oil spills. A resolution cell which contains oil will affect its neighbors to some extent, whereas a piece of noise will only affect one measurement, i.e., one resolution cell. This gives a sequential dimension to the classification problem, which indicates that recurrent networks might be useful, since they could possibly learn to overcome and exploit this stretching effect.

10.3   Experiments

This section will first discuss the ANN architectures used here, and then present a number of experiments and results concerning reliability and robustness of these architectures. The reader who is mostly interested in the general nature of the case study, but not in the details of the experiments and the problem domain at hand, might want to skip the sections covering the experiments or just have a look at the results summarized in tables 10.2 and 10.3, and figures 10.13 to 10.15, and 10.17.

10.3.1   The architectures

We have so far identified that we need to use the intensity and spectral widths information for an individual cell, combined with the same information from neighboring cells. We also identified the stretching effect in the flight direction as a problem as well as a possible source of information which could be useful to identify small oil spills. In this section, we will evaluate the performance of three different architectures; all of them using the same ten input values, i.e., the normalized intensity and spectral width deviations of the resolution cell and its four direct neighbors, and all of them trained to produce the same one output value (i.e., the degree of oil contained in each cell) as obtained from the original environment map(s).

In detail, for the experiments discussed here ANNs of the following three architectures have been used, all of them being trained with the backpropagation algorithm, using a momentum term and an adaptive learning rate, all of them using the logistic activation function for hidden and output units:

- Three-layer feedforward (FF) networks using seven hidden units,
- Recurrent networks similar to those of Jordan with weighted feedback from the output to the hidden layer (see figure 10.8, hereafter referred to as recurrent architecture one (RAI) (in addition to the architecture used here, Jordan's original networks contained a weighted feedback connection from the context units to themselves).

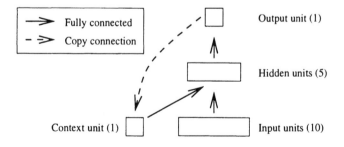

Figure 10.8. Recurrent architecture one (RAI).

- Recurrent networks similar to the simple recurrent network (SRN) proposed by Elman [28], in which the hidden unit activity at time step t is used as additional input at time step $t - 1$ (see figure 10.9, hereafter referred to as recurrent architecture two (RAII) (Elman trained his SRNs to predict the next element in a sequence presented to the network, therefore input and output layer were of the same size. Moreover Elman used different activation functions to us).

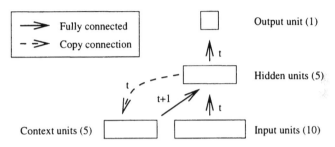

Figure 10.9. Recurrent architecture two (RAII).

For all three architectures initial experiments with different hidden layer sizes were carried out, but did not deliver substantially better results than those documented here.

Both RAI and RAII networks have been trained to classify sequences of resolution cells (including their direct neighbors) in the flight direction individually. So the general approach here is to make use of the context of the previous cell's classification, which is based on the fact that, as illustrated in figure 10.6, it is not possible to measure a clear water echo in one time step and a clear oil echo in the next time step. Instead the transition at the border from water to oil will normally take 4–5 time steps during which oil's 'share' of the returns grows from 0% to 100% (see figure 10.6). Correspondingly, there will be a similar sequence of time steps at the border from oil back to water.

Hence, both recurrent architectures aim at capturing the information contained in the temporal sequence of changes over a few time steps, from water to oil echo and the other way round, and at exploiting the sequential information contained in this transition for the classification task.

10.3.2 Reliability and sensitivity

In addition to environment A, a second environment (B), shown in figure 10.10, has been created especially to evaluate the detection reliability for small oil spills. This environment contains 20 oil spills, ranging in size from very small ones (minimum 20 m × 20 m), which should be undetectable due to the fact that they are (much) smaller than a single resolution cell (20 m × 75 m), to larger ones up to 120 m × 80 m, which should be detected relatively easily.

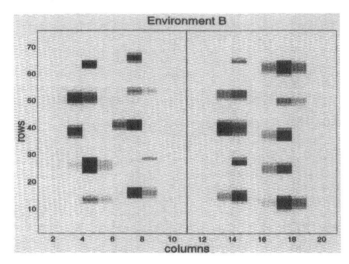

Figure 10.10. Environment B.

The measurements of intensity and spectrum width within this environment (in sea state 3) are shown in figure 10.11. Similar observations can be made here as earlier for environment A: the larger oil spills appear relatively clearly (at least in the intensity map) whereas the smaller ones (seem to) disappear in the background noise.

The resolution cell vectors (see figure 10.12) for this environment, however, show a slightly different picture from those of environment A (see figure 10.7). As there are no really large oil spills in environment B, the cluster of cells easily classifiable as oil is not present here. Instead the vectors now almost form one single cluster, but it can be seen that the oil cells, although mostly within this cluster, tend to be in the lower left area.

To evaluate the different architectures' performance on detecting small oil spills, 14 networks of each architecture have been trained on one half of environment B, seven of them using the left half, the other seven using the right half as training set. The experiments described here were carried out for one particular sea state (3), but similar results were obtained when using other sea states. All networks have been trained to a point where in initial experiments they achieved best performance when tested on the whole environment. For FF networks that was 500 epochs, for both RAI and RAII 1000 epochs. The performance of the ten best networks of each class when tested on all of environments A and B is summarized in table 10.2. The results clearly show that, although being trained on environment B, all networks show better performance in the simpler environment A. RAII networks show slightly better performance than those of RAI, the FF networks' performance is significantly worse.

To further illustrate the above results, the best (continuous) output maps for each architecture are shown in figures 10.13, 10.14 and 10.15.

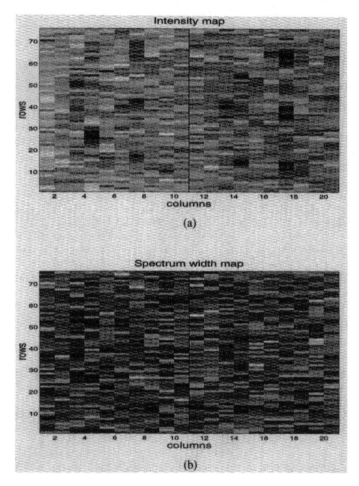

Figure 10.11. Measurements of (a) intensity and (b) spectrum width in environment B (sea state 3).

Table 10.2. Performance of ANNs trained on 50% of environment B, tested on all of environments A and B (sea state 3 only).

ANN type	Detected (B)	False (B)	Detected (A)	False (A)
FF (10)	112 (56%)	6	65 (93%)	9
RAI (10)	124 (62%)	4	69 (99%)	2
RAII (10)	132 (66%)	1	70 (100%)	0

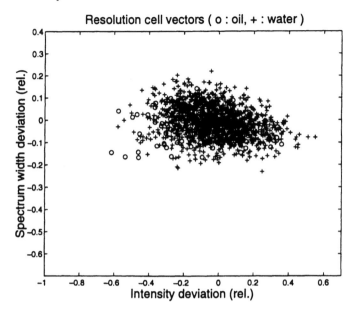

Figure 10.12. Normalized resolution cell vectors for environment B and their binary classification.

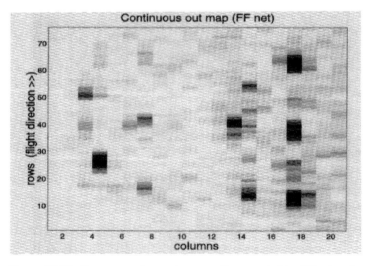

Figure 10.13. Output map for environment B, generated by FF network.

As expected, these maps show that it is the smallest oil spills which cannot be detected. Nevertheless, using RAII, the oil spill detection works rather reliably down to a size of about 50 m × 50 m. Regarding the relatively low resolution of the radar used here this result can be considered fully satisfactory.

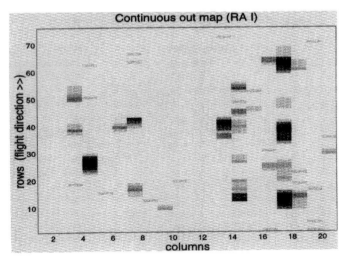

Figure 10.14. Output map for environment B, generated by RAI network.

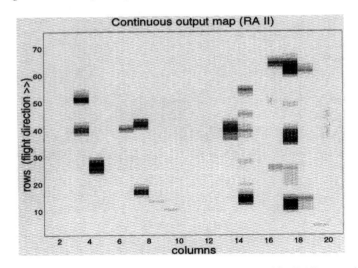

Figure 10.15. Output map for environment B, generated by RAII network.

10.3.3   Robustness to varying sea states

For an evaluation of the ANNs' robustness to varying wind/weather conditions data from the six sea states shown in table 10.1 were used in the experiments discussed here.

In all sea states the normalized relative deviations in intensity and spectrum width show a similar distribution (into two clusters plus an overlap area) as illustrated earlier for environment A in sea state 3 (see figure 10.7), but, as illustrated in figure 10.16, in lower sea states (with less wind and lower waves,

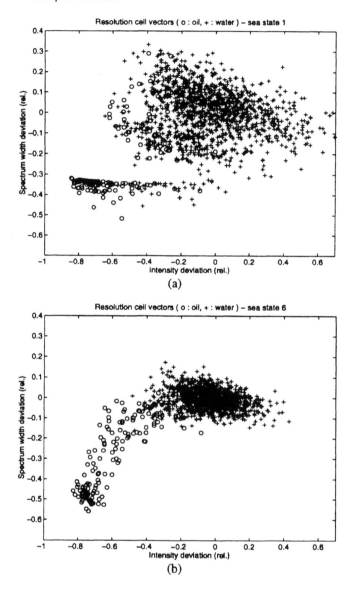

Figure 10.16. Resolution cell vectors for environment A in (a) sea state 1 and (b) sea state 6.

and therefore less clear differences between oil and water surfaces) the values spread out more whereas in higher sea states the clustering is even stronger.

To evaluate the ANNs' robustness to variations in sea state, five networks of each architecture were trained in the same way as described above, but this time on both the left halves of environments A and B, both in sea state 3. These networks were tested on the whole of both environments in all sea states. It has, however, turned out that the results obtained for sea state 1 were dramatically worse than those for other sea states. This can be explained by the fact that for very low waves (as in sea state 1) the rather small differences between oil and water can easily 'disappear' in the noise. Therefore the results for sea state 1 have been omitted here; the results for the remaining five sea states (2–6) are shown in table 10.3. Note that the maximum number of detectable oil spills is 500 (5 networks × 5 sea states × 20 oil spills) in environment B, and 175 (5 × 5 × 7) in environment A. It also turned out that more than 50% of the false alarms occurred in sea state 2, therefore the numbers of false alarms are stated with and without (w/o) state 2 in table 10.3.

Table 10.3. Performance of ANNs trained on half of environments A and B (sea state 3), tested on all of both environments in sea states 2–6.

ANN type	Detected (B)	False (B) (w/o state 2)	Detected (A)	False (A) (w/o state 2)
FF (5)	254 (51%)	31 (13)	163 (93%)	12 (6)
RAI (5)	284 (57%)	23 (9)	171 (98%)	4 (1)
RAII (5)	322 (64%)	8 (2)	175 (100%)	0

It can be seen that, although the networks have only been trained for one particular sea state, the classification performance across five sea states (2–6) is not much worse than that earlier for sea state 3 alone. Again, the RAII networks clearly exhibit the best performance. Figure 10.17 shows the continuous output values computed by one of these networks for environment A (in sea state 3) as well as the resulting binary classification (using an output threshold of 0.5). When comparing these maps to the correct environment map (originally shown in figure 10.4) it can be seen that the output as generated by an RAII network actually comes very close to the original.

10.3.4   Further experiments

All experiments discussed here so far were based on sea environments only containing water and oil, and measurements from an illumination distance of 20 km. In practice, however, sea environments do of course also contain islands and boats, and illumination distances differ, especially when surveying large areas. Hence the capacity to generalize across different illumination distances as well as different object types is very important.

In [110] RAII networks were trained on data from an illumination distance of 20 km (sea state 3 only), and tested on data obtained from measurements from

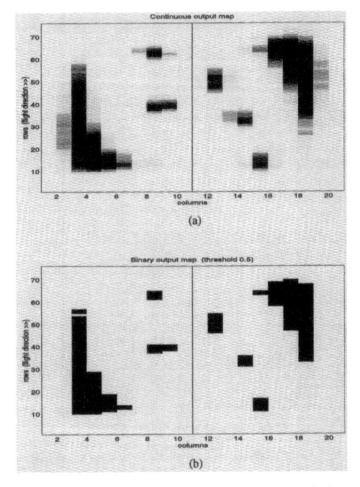

Figure 10.17. Continuous output map for environment A (sea state 3) from an RAII network, and resulting binary classification.

illumination distances of 10 and 40 km in all sea states. A halving/doubling of the illumination distance also leads to a halving/doubling of the (relevant) beam width. Hence, the images obtained from a shorter distance are clearer and the detection/segmentation becomes easier, and correspondingly for a longer illumination distance the detection of small oil spills becomes more difficult. Accordingly very good results were achieved for the 10 km distance. Results for the 40 km distance are significantly worse, but still satisfactory for higher sea states (for sea states 3 to 6 most of the networks, although being trained on 20 km distance data only, detected three or four of the five small oil spills while not generating any false alarm).

In [111] experiments have been extended to sea environments including islands and boats in addition to water and oil spills. Best results were obtained with a more complex recurrent architecture (using second-order recurrent connections), but good performance was also obtained with RAII networks, as they have been discussed here.

10.4 Analysis and discussion

As hinted earlier the selection of network architecture introduces both constraints and possibilities for learning. The most important variable aspect of the documented experiments is the information available to the classifier. The FF network can only account for the data for the current cell and the corresponding neighboring cells. It has already been argued that there is a probable dependence between the classification of cells along the flight direction. A strategy sensitive to this stretching effect has at least two advantages:

- The system gets more robust, less sensitive to sudden fluctuations imposed by e.g., noise.
- Reliable classification of large oil spills (or large areas of water).

RAI has the previous classification result as part of its input. Thus, it is quite natural to believe that the previous classification affects the result for the current cell. Still, the learning process has to detect the importance and (possibly) necessity of such information. So far, no substantial evidence has been put forward that such features actually are used.

RAII uses the hidden activity of the previous classification (again in the flight direction) as part of its input. Thus, a certain influence and stretching effect on the classification was expected. Similar to RAI, the importance of this influence has to be detected by learning from the examples. Learning in RAII only affects the development of hidden activity for the current classification, i.e., at time step t, which may make use of the context units, i.e., the features developed for the previous classification. The learning in RAII, however, cannot influence the activity of the context units, i.e., the hidden unit activity at time step $t - 1$, but only change the weights from the context units to the hidden layer.

It is far from clear what is actually captured by these hidden states and what makes this architecture perform better than RAI (and FF networks).

The analysis is divided into two parts. First, functional problems that the three architectures have are mapped out. This analysis points out particular weaknesses of each architecture in terms of the concepts appearing in our particular domain (i.e., oil spills and water). Second, general tendencies for the internal representations developed in each architecture are investigated. Through analysis of hidden activity the arguments for the performance of the networks can be substantiated in terms of the network primitives (i.e., units and weights).

Again, the reader not interested in the detailed analyses might want to skip the following two subsections, and read the brief summary which is part of the concluding section.

10.4.1 Functional analysis

Learning optimizes weights in the network according to a *global* error measure involving the complete set of training examples. Thus, it is likely that the weights implement *individual* mappings with varying success. To investigate this further the actual output of networks of each architecture (trained on the left half of environment A only) was recorded for two flight directions (the one used for training and previous testing, and the opposite) in environment A. The outputs of the two flight directions were compared for each resolution cell in the environment (see figure 10.18). A misfit is either negative (black) or positive (white) depending on which flight direction gives rise to the most distinct response. If a cell is classified the same from both directions it is shown gray in figure 10.18. Note that does not necessarily mean that the classification is actually correct.

The FF network generally makes mistakes whenever the classified cell is surrounded by cells of opposite types (for example, in figure 10.18 cells (2, 22–35), (4, 26–38) and (17, 47–53)). It is the left and right neighbor cells that have negative influence. In addition, the effect is different for classification on the right side and the left side of the oil spill (negative difference on one side and positive difference on the other). This may be due to the limited training set used. On the other hand, it seems less inclined to make mistakes along the flight direction. It does not seem to matter in which direction oil spills are found. In general the noise tolerance is not good. Different flight directions sometimes result in different classifications for the same distorted radar echo. The difference in noise sensitivity is particularly evident if we compare the image for the FF network with the image for RAI.

The RAI map (see figure 10.18) is much smoother and thus reveals better robustness to noise. Apart from better noise tolerance, RAI shares many of the problems noted for FF. For example, the influence of the neighboring cells distracts classification of cells which lie beside oil spills (e.g., cells appearing at coordinates (4, 27–32), (17, 47–52) and (19, 47–53)). Here too, it makes a difference if the oil spill appears on the right or left side. So, the additional input with the previous classification inhibits the introduction of noise but does not disable distinct, but incorrect, classifications similar to those made by FF.

In RAII much less can be foreseen by just looking at the architecture. However, looking at the differences between the two flight directions reveals quite a lot. First, noise is even less detectable. The influence of the prior hidden state at the input inhibits noise. Second, a stretching effect appears distinctively, both in terms of supporting neighboring cells and the middle cell. The bad influence from the neighbors on the left and right side is less effective

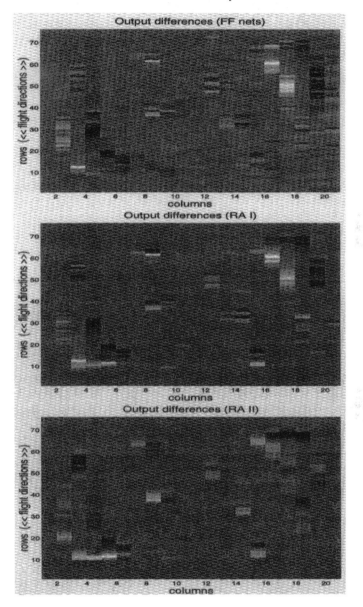

Figure 10.18. For each cell the difference between the output when flying in the normal direction and the output when flying in the opposite direction is shown. Crudely put, gray is used when both runs agree on the same output, whereas white and black represent positive and negative differences caused by the different flight directions.

since their contribution is seemingly made less effective by the possibility of learning to exploit the stretching effect. Thus, it seems that the classification of a cell is less sensitive to its neighbors when the hidden activity of the previous classification is available. The bad news is that this makes transitions between oil and water along the flight direction less clear. It is a clear difference between the outputs of the network when an oil spill is entered from two directions (e.g., cells with coordinates (3–5, 10–12) (3, 53–58) and (16–18, 66–68)) evident in the white areas below and black areas above oil spills. It takes more time-steps to detect the oil spill and it takes more steps to detect that it ends. Strangely, this does not appear very clearly in RAI. One possible explanation would be that the RAI has less possibility of using the delay and therefore (to optimize its overall performance) relies more on the neighboring cells. RAII can, on the other hand, use the full power of the delay and thus loses the precision of detecting the smooth borders. One possible improvement could be through the use of some extended learning technique (e.g., weights on the recurrent connections in the RAII), since that would allow the learning of classifying a cell to develop features of the previous classification.

To sum up the evidence from this functional analysis, there are two main points. First, there is a stretching effect. The classification of the cells in RAII relies on the previous classification (and the one before that etc). In RAI it is less obvious that the history (which only can go one step back) does anything else than reducing the impact of small fluctuations. Second, there is a harmful sensitivity to the left and right neighboring cells. To manage without the stretching effect such data are vital to manage even the simplest cases. However, in the documented experiments this usefulness seems to be over-learned (e.g., different sensitivity to right and left). RAII compensates this influence fairly well with the delayed input and thus seems to offer the most reliable architecture in spite of the differences between 'entering' and 'leaving' oil spills.

10.4.2 Analysis of internal states

So far what is captured by these internal states (taking the form of hidden activity) has only been analyzed indirectly. In this section a hierarchical cluster analysis (HCA) is performed on the hidden activities developed for each cell, for the FF and RAII architectures. This way, the information captured by the internal states is made available to some extent. As with most analysis techniques some information is lost. Objections have been raised against the approach of relying on spatial relations (in a Euclidean space) for describing what these states mean since it does not take into account the weights which the states are subjected to ([207]; see also Niklasson and Bodén, chapter 2 of [1]). However, without making any claims about the content and properties of such states, HCA provides a good measure of what is (nevertheless crudely) captured by these states.

We will now compare the organization of these states for the two types of network in terms of what they are composed from (the input) and what they give rise to (the output). To make this procedure illustrative, cluster dendrograms for both the FF and RAII architectures were generated for a representative subset of cells in environment A (see figures 10.19 and 10.20). Columns 15 and 18 were used and some cells in areas of no specific interest were removed (in 15 cells 20–55 and in 18 cells 1–15). The RAI is not shown since the FF and RAII dendrograms are sufficient to prove our point. Each entry in the diagrams corresponds to a cell in the map and is labeled according to the cell type (W for 'pure water', O for cells containing any amount of oil), its position in the map in brackets, and the oil quantity used for the SLAR simulation model (between 0.0 and 1.0, see figure 10.4). The Euclidean distance between two states (or state-clusters) is indicated along the x-axis.

For the FF architecture there are three main clusters present in the cluster analysis dendrogram shown in figure 10.19. The three main clusters crudely correspond to 'mainly water', 'just oil' and 'borderline cases'. There is also a large number of scattered entries which are mainly either borderline cases or misclassifications. The 'mainly water' cluster contains entries for the cells contained in large water areas, with a few exceptions. As it turns out these exceptions are misclassified in the FF experiment. There is also a large number of water cells not contained in this cluster. The majority of those have oil cells as neighbors. We prioritized columns which crossed oil spills and not those that were parallel to them. Thus, no further evidence of harmful influence from neighboring cells is demonstrated in the diagrams. However, a number of vague cases appears scattered in the dendrogram. The 'just oil' cluster contains almost without exceptions the clear oil cases (oil level is above 0.6). Those not contained are either vague cases (with close to no oil) or borderline cases. There is a main 'borderline cases' cluster. There is no other obvious relation between the entries. Overall, little structure is found in the dendrogram for the FF architecture.

The case is different for RAII shown in figure 10.20. Again, three main clusters are available. One contains clear cases of oil, one contains clear cases of water and the third contains the transition cases. Overall there is a clear tendency of distinguishing between normal (appears in dense clusters) and more or less obscure cases (scattered or loosely clustered). There is a fourth cluster (in the bottom) which only covers transitions from oil to water, containing both oil and water cells. There are also smaller clusters specific to particular oil spills (see e.g., the lower, small oil spill in column 15 and the oil spill in the top of column 18). This tendency corresponds closely to the problem of distinguishing when water becomes oil spill and *vice versa*. The states are closely related spatially and it is therefore easy to see the problem for the classifier to distinguish between them. It also indicates how transitions are implemented. Rather than jumping between completely different states in hidden space, there are smooth transitions in hidden space leading to 'attractors' (the large, dense clusters). This is different from the FF case for which hidden states either correspond to a finite set of states (the clear cases) the classifier recognizes, or some less 'meaningful' states which

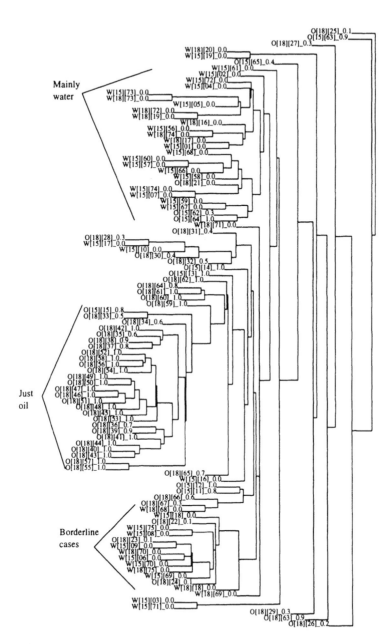

Figure 10.19. A hierarchical cluster dendrogram generated from a subset of the available hidden states (representing particular cells) which were developed as a result of learning in the FF architecture.

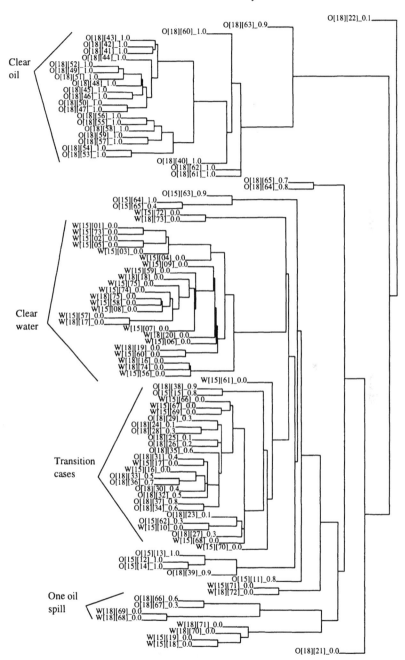

Figure 10.20. A hierarchical cluster dendrogram generated from a subset of the available hidden states (same as for FF) which were developed as a result of learning in the RAII architecture.

are misclassified. Furthermore, in the FF network there are no transitions. Thus, there is no point (for the learning) in finding relations between 'states'.

10.5   Summary and conclusion

The intensity and spectrum width values extracted from each resolution cell's Doppler spectrum have been identified as the relevant features to be used as input to the classification networks. The need to normalize these values with regard to the sea state specific reference values in order to allow generalization across different sea states was discussed. Furthermore, it was shown that reliable classification of resolution cells on the basis of their own values only is not possible, such that the values of their neighbor cells should be used as additional input. A discussion of the stretching effect due to overlapping radar echoes indicated that recurrent networks might be useful to compensate and exploit the sequential nature of this effect.

An evaluation of three different ANN architectures, one feedforward and two recurrent architectures, regarding the reliability/sensitivity of the detection of small oil spills as well as the robustness to varying sea states shows that significantly better performance is exhibited by the recurrent architectures. In particular, best performance is achieved by RAII networks, which are architecturally similar to Elman's [28] SRNs. It has been shown that networks of this recurrent architecture have been successfully trained such that:

- They have a high detection reliability and sensitivity, even to relatively small oil spills.
- The number of false alarms could be kept very low, i.e., a low sensitivity to noise has been achieved.
- The networks are very robust to variations in sea state and illumination distances (the latter has not been discussed here in detail) for sea states 3 to 6 (and, with some limitations, 2).

The computational costs of the technique presented here are rather low as the only pre-processing required is the calculation of average intensity and spectrum width per resolution cell, and no complete radar images are required, i.e., the presented technique allows a real-time detection of oil spills.

The observed performance relies on several findings. From the functional analysis it was noted that learning in the FF and RAI architectures optimized classification performance by relying on the neighboring cells. This led to distinct classifications in the transition regions. However, the classifications for cells parallel to oil spills were sometimes incorrect. This may be explained by the limited training set. The noise tolerance was increased by the use of recurrent connections in which activity from prior classifications affected the performance. The activity of previous classifications compensated for the sometimes harmful influence from the neighboring cells in RAII. However, the memory trace of prior mappings made the classification of the borders along the flight direction less distinct.

Further analysis on the internal states developed in the networks provided possible explanations to these findings. First, the RAII network develops states for cells reflecting their context to a high degree. Through cluster analysis this was evident since similar situations were densely clustered and borderline cases were positioned in state space to provide smooth transitions between the oil and water clusters. The FF network with no recurrent connections could not make use of these dynamic transitions. Instead the state space for the FF architecture contained less structure and many cases of hidden states for cells were scattered in state space. Thus, the feedforward network provided poor generalization for the domain at hand. The recurrent networks (RAII in particular) can learn to make use of the stretching effect and thus generalize to a higher degree.

Hence, the results reported in this paper can be considered very satisfactory. The degree of robustness and generalization achieved here across a large variety of simulated sea conditions is very promising for an application of this method under real world conditions. The approach taken here, to have an ANN generate a rather clear output map from two noisy input maps, is a rather general one as it should be applicable to a number of other pattern recognition problems.

Acknowledgments

The work presented in this chapter is part of a long-term joint project on ANNs for radar signal processing between the Connectionist Research Group (CRG), University of Skövde, and Ericsson Microwave Systems, Mölndal, Sweden. The CRG's work on this project is currently being supported by PREEMs Miljöstiftelse, and between 1994 and 1996 partial funding was supplied by the Swedish Ministry of Education and Science.

Bibliography

[1] A. Browne *Neural Network Perspectives on Cognition and Autonomous Robotics* Institute of Physics Publishing, Bristol, UK, 1997.

[2] N. Ahmed, T. Natarajan and K. R. Rao. Discrete cosine transform. *IEEE Transactions on Computers*, C-23(1):90–93, January 1974.

[3] A. Andronov. *Theory of Oscillators (Translated from Russian by F. Immirzi).* Pergamon, London, 1966.

[4] R. Ash. *Information Theory*. Wiley, New York, 1965.

[5] J. Backman, M. Hebert and M. Seidenberg. Aquisition and use of spelling–sound information in reading. *Journal of Experimental Child Psychology*, 38:114–133, 1984.

[6] A. G. Barto. Neuron-like adaptive elements that can solve difficult learning control problems. *IEEE Transactions on Systems, Man and Cybernetics*, (SM-C-13):834–846, 1983.

[7] M. Brown and C. J. Harris. *Neurofuzzy Adaptive Modelling and Control*, Prentice-Hall, Hemel Hempstead, 1994.

[8] J. A. Bullinaria. Internal representations of a connectionist model of reading aloud. In *Proceedings of the 16th Annual Conference of the Cognitive Science Society*, pages 84–89, Erlbaum, Hillsdale, NJ, 1994.

[9] J. A. Bullinaria. Representation, learning, generalisation and damage in neural network models of reading aloud. *Technical Report*, Edinburgh Univerity, 1994.

[10] J. A. Bullinaria. Modelling reaction times. In L. S. Smith and P. J. B. Hancock, editors, *Neural Computation and Psychology*, pages 34–48. Springer, New York, 1995.

[11] J. A. Bullinaria. Neural network learning from ambiguous training data. *Connection Science*, 7:99–122, 1995.

[12] J. A. Bullinaria and N. Chater. Connectionist modelling: Implications for cognitive neuropsychology. *Language and Cognitive Processes*, 10:227–264, 1995.

[13] De-Fu Cai and Ming Zhou. Adaptive image compression based on back-propagation neural networks. In *Proceedings of the SPIE - volume 1766*, pages 678–683. SPIE, Bellingham, WA, 1992.

[14] G. A. Carpenter and S. Grossberg. The art of adaptive pattern recognition by a self-organizing neural network. *Computer*, pages 77–88, March 1988.

[15] Oscal T.-C. Chen, Bing J. Sheu and Wai-Chi Fang. Image compression using self-organization networks. *IEEE Transactions on Circuits and Systems for Video Technology*, 4(5):480–489, Oct 1994.

[16] L. O. Chua. *IEEE Transactions on Circuits and Systems*, 13:815–826, 1993.

[17] S. Churcher. *VLSI neural networks for computer vision.* Unpublished PhD dissertation, Department of Electrical Engineering, Edinburgh University, March 1993.

[18] M. Cohen and S. Grossberg. *IEEE Transactions on Systems, Man and Cybernetics*, 13:815–826, 1983.

[19] M. Coltheart, B. Curtis, P. Atkins and M. Haller. Models of reading aloud: Dual-route and parallel distributed processing approaches. *Psychological Review*, 100:589–608, 1993.

[20] C. Cramer, E. Gelenbe and H. Bakircioglu. Low bit rate video compression with neural networks and temporal subsampling. *Proceedings of the IEEE*, 84(10):1529–1543, October 1996.

[21] J. M. Cruz and L. O. Chua. IEEE transactions on circuits and systems. *IEEE Transactions on Circuits and Systems*, 38:812–817, 1991.

[22] J. G. Daugman. Complete discrete 2-D Gabor transforms by neural networks for image analysis and compression. *IEEE Transactions on Acoustics, Speech and Signal Processing*, 36(7):1169–1179, July 1988.

[23] J. G. Daugman. Relaxation neural network for non-orthogonal image transforms. In *IEEE International Conference on Neural Networks*, pages 547–560. IEEE, New York, July 1988.

[24] de Savigny M and Newcomb R W. Realization of Boolean functions using a pulse coded neuron. In Zaghloul M E, Meador J L and Newcomb R W, editors, *Silicon Implementation of Pulse Coded Neural Networks*, pages 65–77. Kluwer, Boston, MA, 1994.

[25] P. Devijer and J. Kittler. *Statistical Pattern Recognition.* Prentice-Hall, Englewood Cliffs, NJ, 1982.

[26] De Yong M and Fields C. Silicon neurons for phase and frequency detection and pattern generation. In M. E. Zaghloul, J. L. Meador and R. W. Newcomb, editors, *Silicon Implementation of Pulse Coded Neural Networks*, pages 65–77. Kluwer, Boston, MA, 1994.

[27] J. G. Elias. In M. E. Zaghloul, J. L. Meador and R. W. Newcomb, editors, *Silicon Implementation of Pulse Coded Neural Networks*, pages 39–63. Kluwer, Boston, MA, 1994.

[28] J. L. Elman. Finding structure in time. *Cognitive Science*, 14:179–211, 1990.

[29] J. L. Elman. Learning and development in neural networks: The importance of starting small. *Cognition*, 48:71–99, 1993.

[30] B. Everitt. *Cluster Analysis.* Wiley, New York, 1980.

[31] Wai-Chi Fang, Bing J. Sheu and Oscal T.-C. Chen. A real-time VLSI neuroprocessor for adaptive image compression based upon frequency-sensitive competitive learning. In *IJCNN-91-Seattle Part 1 (of 2)*, pages 429–435. IEEE, New York, July 1991.

[32] Wai-Chi Fang, Bing J. Sheu, Oscal T.-C. Chen and Joongho Choi. A VLSI neural processor for image data compression using self-organization networks. *IEEE Transactions on Neural Networks*, 3(3):506–518, May 1992.

[33] P. Gallinari, S. Thiria, F. Badran and F. Fogelman-Soulie. On the relations between discriminant analysis and multilayer perceptrons. *Neural Networks*, 4:349–360, 1991.

[34] E. Gelenbe. Random neural networks with negative and positive signals and product form solution. *Neural Computation*, 1(4):502–511, 1989.

[35] E. Gelenbe. Stability of the random neural network model. *Neural Computation*, 2(2):239–247, 1990.

[36] E. Gelenbe. Learning in the recurrent random neural network. *Neural Computation*, 5(1):154–164, 1993.

[37] E. Gelenbe and A. Stafylopatis. Global behaviour of homogeneous random neural systems. *Applied Mathematical Modelling*, 15:534–541, 1991.

[38] E. Gelenbe and M. Sungur. Random network learning and image compression. In *Proceedings of the 1994 IEEE International Conference on Neural Networks. Part 6 (of 7)*, pages 3996–3999, IEEE, Piscataway, NJ, 1994.

[39] E. Gelenbe, M. Sungur and C. Cramer. Learning random networks for compression of still and moving images. In *A Decade of Neural Networks—a Workshop at the Jet Propulsion Laboratory*, pages 171–189. JPL, May 1994.

[40] S. Grossberg. Self Organizing networks. In E. L. Schwartz, editor, *Computational Neuroscience*, page 63. MIT Press, San Mateo, CA, 1990.

[41] S. Grossberg. Adaptive pattern classification and universal recoding: Parallel development and coding of neural feature detectors. *Biological Cybernetics*, 23:121–134, 1976.

[42] A. Hamilton, A. F. Murray, D. J. Baxter, S. Churcher, H. M. Reekie, and L. Tarassenko. Integrated pulse stream neural networks: results, issues and pointers. *IEEE Transactions on Neural Networks*, 3(3):385–393, 1992.

[43] M. J.R. Healy. *Matrices for Statistics*. Clarendon, Oxford, 1986.

[44] M. Hirsch. *Siam Journal of Mathematical Analysis*, 16:423–439, 1985.

[45] J. Hopfield. *Proceedings of the National Academy of Sciences*, 81:3088–3092, 1984

[46] K. Hornick. *Neural Networks*, 2(5):359–366, 1989.

[47] G. Jackson and A. F. Murray. Competence acquisition in an autonomous mobile robot using hardware neural techniques. In D. Touretzky, M. Mozer and M. Hasselmo, editors, *Advances in Neural Information Processing Systems*, number 8, pages 1031–1037, MIT Press, Cambridge, MA, 1996.

[48] T. A. Johansen. In *Proceedings Fourth IEEE International Conference on Fuzzy Systems*, pages 97–102, IEEE, New York, 1995.

[49] M. Joy. *Annals of Mathematical and Machine Intelligence*, 2:131–158, 1995.

[50] R. Kamimura and S. Nakanishi. Hidden information maximization for feature detection and rule discovery. *Network: Computation in Neural Systems*, 6:577–602, 1995.

[51] R. Kohno, M. Arai and H. Imai. Image compression using a neural network with learning capability of variable function of the neural unit. In *Visual Communications and Image Processing '90*, pages 69–75, Lausanne, October 1990. International Society for Optical Engineering, Bellingham, WA, 1990.

[52] T. Kohonen. Springer, New York, 1988.

[53] T. Kohonen. *Self-Organization and Associative Memory*. Springer, Berlin, 1984.

[54] B. Kosko. *Neural Networks and Fuzzy Systems: A Dynamical Systems Approach to Machine Intelligence*. Prentice-Hall, Englewood Cliffs, NJ, 1992.

[55] W. J. Krzanowski and D. Partridge. Software diversity: Practical statistics for its measurement and exploitation. *Research Report 324*, University of Exeter, Computer Science Department, 1995.

[56] S. Kuffler. *From Neuron to Brain*. Freeman, San Francisco, CA, 1984.

[57] P. Ladefoged. *A Course in Phonetics*. Harcourt Brace, 1993.

[58] W. P. Lincoln and J. Skrzpek. Synergy of clustering multiple backpropagation networks. In D. S. Touretzky, editor, *Advances in Neural Information Processing Systems 2*, pages 650–657, Morgan Kaufmann, San Mateo, CA, 1990.

[59] Y. Linde, A. Buzo and R. M. Gray. An algorithm for vector quantization. *IEEE Transactions on Communications*, COM-28(1):84–95, January 1980.

[60] B. Littlewood and D. R. Miller. Conceptual modelling of coincident failure in multi-version software engineering. *IEEE Transactions on Software Engineering*, 15(12):1596–1614, 1989.

[61] Cheng-Chang Lu and Yong Ho Shin. A neural network based image compression system. *IEEE Transactions on Consumer Electronics*, 38(1):25–29, Feb 1992.

[62] S. Marsi, G. Ramponi and G. L. Sicuranza. Improved neural structures for image compression. In *Proceedings, International Conference on Acoustic Speech and Signal Processing (Toronto, Canada, May 1991)*, pages 2821–2824, IEEE, Piscataway, NJ, 1991.

[63] J. L. Meador and P. D. Hylander Pulse coded winner-take-all networks. In M. E. Zaghloul, J. L. Meador and R. W. Newcomb, editors, *Silicon Implementation of Pulse Coded Neural Networks*, pages 79–99. Kluwer, Boston, MA, 1994.

[64] J. L. Meador and P. D. Hylander *Silicon Implementation of Pulse Coded Neural Networks*. Kluwer, Boston, MA, 1994.

[65] W. T. Miller. *Neural Networks for Control*. MIT Press, Cambridge, MA, 1990.

[66] M. C. Mozer and P. Smolensky. Using relevance to reduce network size automatically. *Connection Science*, 1(1):3–16, 1989.

[67] N. M. Nasrabadi and Y. Feng. Vector quantization of images based upon Kohonen self organizing feature maps. In *Proceedings, International Conference on Neural Networks*, pages 101–108, San Diego, CA, IEEE, New York, July 1988.

[68] H. Nema. Phonotactis and sonority (in Japanese). *Senshu Journal of Foreign Language and Education*, pages 65–94, 1994.

[69] R. Hecht-Nielsen. Counterpropagation networks. *Applied Optics*, 26:4979–4984, 1987.

[70] S. J. Nowlan and G. E. Hinton. Evaluation of adaptive mixtures of competing experts. In R. P. Lippmann, J. S. Moody and D. S. Touretzky, editors, *Advances in Neural Information Processing Systems 3*, pages 774–780. Morgan Kaufmann, San Mateo, CA, 1991.

[71] J. Palis and S. Smale. In *Proceedings of the Symposium in Pure Mathematics*, volume 14, RI, 1970. American Mathematics Society.

[72] K. Papathanasiou and A. Hamilton. Pulse based signal processing: VLSI implementation of a Palmo filter. In *Proceedings of the International Symposium on Circuits and Systems*, volume 1, pages 270–273, Atlanta, May 1996.

[73] D. Partridge. On the difficulty of really considering a radical novelty. *Minds and Machines*, 5:391–410, 1995.

[74] D. Partridge. Network generalisation differences quantified. *Neural Networks*, 9(9):263–271, 1996.

[75] D. Partridge and W. B. Yates. The replicability of neural computing experiments. *Research Report 305*, University of Exeter, Department of Computer Science, 1995.

[76] D. Partridge and W. B. Yates. Engineering multiversion neural-net systems. *Neural Computation*, 8(4):869–893, 1996.

[77] G. Qiu, M. R. Varley and T. J. Terrell. Image compression by edge pattern learning using multilayer perceptrons. *Electronic Letters*, 29(7):601–603, April 1993.

[78] S. A. Rizvi, Lin-Cheng Wang and N. M. Nasrabadi. Neural network vector predictors with application to image coding. In *Proceedings of the 1995 IEEE International Conference on Image Processing. Part 3 (of 3)*, pages 296–299. IEEE, New York, 1995.

[79] T. Rosaka and J. Hamori. Neural implementations. *IEEE Transactions on Circuits and Systems*, 40:182–195, 1993.

[80] D. E. Rumelhart, G. E. Hinton and R. J. Williams. *Learning Internal Representations by Error Propagation*, volume 1. MIT Press, Cambridge, MA, 1986.

[81] G. E. Hinton, D. E. Rumelhart and R. J. Williams. Learning internal representations by error propagation. In *Parallel Distributed Processing*, pages 533–536. Morgan Kauffman, San Mateo, CA, 1986.

[82] T. R. Schultz and J. L. Elman. Analyzing cross connected networks. In G. Tesauro J. D. Cowan and J. Alspector, editors, *Advances in Neural Information Processing Systems*, volume 6, pages 1117–1124. Morgan Kaufmann, San Mateo, CA, 1994.

[83] M. S. Seidenberg and J. L. McClelland. A distributed, developmental model of word recognition and naming. *Psychological Review*, 96:523–568, 1989.

[84] R. Setiono and Guojun Lu. Image compression using a feedforward neural network. In *Proceedings of the 1994 IEEE International Conference on Neural Networks (June 1994)*, pages 4761–4765. IEEE, New York, 1994.

[85] Bing J. Sheu and Wai-Chi Fang. Real-time high-ratio image compression using adaptive VLSI neuroprocessors. In *Proceedings—ICASSP*, volume 2, pages 1173–1176. IEEE, New York, 1991.

[86] G. L. Sicuranza, G. Ramponi and S. Marsi. Artificial neural network for image compression. *Electronics Letters*, 26(3):477–479, March 1990.

[87] F. Silva and L. Almeida. Acceleration techniques for the back-propagation algorithm. In *Lecture Notes in Computer Science*, volume 412, pages 110–119. Springer, Berlin, 1990.

[88] C. A. Skarda and W. Freeman. Synaptic potentials. *Behavioral and Brain Sciences*, 10:161–195, 1987.

[89] S. S. Skrzypkowiak and V. K. Jain. Video motion estimation using a neural network. In *Proceedings of the 1994 IEEE International Symposium on Circuits and Systems. Part 3 (of 6)*, pages 217–220. IEEE, New York, 1994.

[90] S. S. Skrzypkowiak and V. K. Jain. Formative motion estimation for translational shear and zoom sequences. In *Proceedings of the 1996 IEEE International Conference on Acoustics, Speech and Signal Processing, ICASSP. Part 4 (of 6)*, pages 1922–1925. IEEE, New York, 1996.

[91] S. A. Solla, E. Levin and M. Fleisher. Accelerated learning in layered neural networks. *Complex Systems*, 2:625–639, 1988.

[92] N. Sonehara, M. Kawato, S. Miyake and K. Nakane. Image data compression using a neural network model. In *Proceedings, International Joint Conference on Neural Networks (Washington, DC, June 1989)*, pages 35–41, IEEE: Piscataway, NJ, 1989.

[93] M. Sugeno and G. T. Kang. Neurofuzzy control. *Fuzzy Sets and Systems*, 26:15–33, 1988.

[94] T. Takagi and M. Sugeno. *IEEE Transactions on Systems, Man and Cybernetics*, 15:116–132, 1985.

[95] J. Tomberg.   Synchronous pulse density modulation in neural network implementation. In M. E. Zaghloul, J. L. Meador and R. W. Newcomb, editors, *Silicon Implementation of Pulse Coded Neural Networks*, pages 65–77. Kluwer, Boston, MA, 1994.

[96] G. Towell and J. Shavlik. The extraction of refined rules from knowledge based neural networks. *Machine Learning*, 131:71–101, 1993.

[97] A. K. Rigler, W. T. Zink, T. P. Vogl, J. K. Mangis and D. L. Alkon. Accelerating the convergence of the back-propagation method. *Biological Cybernetics*, 59:257–263, 1988.

[98] A. van der Sluis and H. A. van der Horst. The rate of convergence of conjugate gradient. *Numererische Mathematik*, 48:543–560, 1986.

[99] H. L. Viktor, A. P. Engelbrecht and I. Cloete. Reduction of symbolic rules from artificial neural networks using sensitivity analysis. In *Proceedings of the IEEE International Conference on Neural Networks*, pages 1788–1793, IEEE, New York, 1995.

[100] R. L. Watrous. Learning algorithms for connectionist networks: Applied gradient methods of nonlinear optimization. In M. Caudill and C. Butler, editors, *IEEE First International Conference on Neural Networks (San Diego, 1987)*, volume 2, pages 619–627, Springer, New York, 1987.

[101] Hsien-Chung Wei, Yung-Ching Chang and Jia-Shang Wang.   Kohonen-based structured codebook design for image compression. In *Proceedings of the 1993 IEEE Region 10 Conference on Computer, Communication, Control and Power Engineering. Part 3 (of 5)*, pages 426–429. IEEE, New York, 1993.

[102] A. S. Weigend, D. E. Rumelhart and B. A. Huberman. Generalization by weight-elimination with application to forecasting. In *Neural Information Processing Systems*, volume 4, pages 950–957. Morgan Kaufmann, San Mateo, CA, 1992.

[103] P. Werbos. Beyond regression: New tools for prediction and analysis in the behavioral sciences. *Technical Report*, PhD thesis, Harvard University, 1974.

[104] S. A. Teucolsky, W. H. Press, B. P. Flannery and W. T. Vetterling. *Numerical Recipes in C*. Cambridge University Press, Cambridge, 1988.

[105] B. Widrow and M. E. Hoff. *IRE WESCON Convention Record*. pages 96–104, 1960.

[106] J. Wiles and M. Ollila. Intersecting regions: The key to combinatorial structure in hidden unit space. In J. D. Cowan, S. J. Hanson and C. L. Giles, editors, *Advances in Neural Information Processing Systems*, volume 5, pages 27–33. Morgan Kaufmann, San Mateo, CA, 1993.

[107] P. M. Williams.   A Marquardt algorithm for choosing the step-size in back-propagation learning with conjugate gradients. *Technical Report CSRP-229*, Cognitive Science, University of Sussex, 1992.

[108] S. Solla, Y. Le Cun and I. Kanter. Eigenvalues of covariance matrices: Application to neural network learning. *Physical Review Letters*, 66:2396–2399, 1991.

[109] T. Ziemke.   A connectionist approach to Doppler radar-based detection of oil spills on water. Master's thesis, Department of Computer Science, University of Skövde, 1994.

[110] T. Ziemke. Radar image segmentation using recurrent artificial neural networks. *Pattern Recognition Letters*, 17(4):319–334, April 1996.

[111] T. Ziemke. Radar image segmentation using self-adapting recurrent artificial neural networks. *International Journal of Neural Systems*, 1997. In Press.

[112] J. Von Neumann. In A. W. Burks, editor, *Theory of Self-Reproducing Automata*. University of Illinois Press, Urbana, IL, 1966.

[113] S. Wolfram. Cellular automata. *Physica D*, 10:1–35, 1984.

[114] R. Woodburn, H. M. Reekie and A. F. Murray. Pulse-stream circuits for on-chip learning in analogue VLSI neural networks. In *Proceedings of the IEEE International Symposium on Circuits and Systems*, volume 4, pages 103–106, 1994.

[115] N. Woodcock. In *Proceedings Sixth International Conference on Artificial Intelligence in Engineering*, pages 903–919, 1991.

[116] F. W. Young. *Multidimensional Scaling, History, Theory and Applications*. Erlbaum, Hillsdale, NJ, 1987.

[117] Zhong Zheng, Masayuki Nakajima and Takeshi Agui. Study on image data compression by using neural network. In *Visual Communications and Image Processing '92*, pages 1425–1433. SPIE, Bellingham, WA, 1992.

[118] S. C. Ahalt, F. D. Garber, I. Jouny and A. K. Krishnamurthy. Performance of synthetic neural network classification of noisy radar signals. In D. S. Touretzky, editor, *Advances in Neural Information Processing Systems*, volume 1, pages 281–288, Morgan Kauffmann, San Mateo, CA, 1989.

[119] H. Akaike. On a successive transformation of probability distribution and its application to the analysis of the optimum gradient method. *Annals of the Institute of Statistical Mathematics*, 11:1–17, 1974.

[120] R. Andrews, J. Diederich and A. B. Tickle. Survey and critique of techniques for extracting rules from trained artificial neural networks. *Knowledge Based Systems*, 8(6):373–389, 1995.

[121] M. Aoki. *Introduction to Optimization Techniques*. Macmillan, New York, 1971.

[122] P. L. Bartlett. *Technical Report*, 1996.

[123] R. Battiti. First and second-order methods for learning: Between steepest descent and Newton's method. *Neural Computation*, 4:141–166, 1992.

[124] R. Battiti and F. Masulli. Bfgs optimization for faster and automated supervized learning. In *International Neural Network Conference*, volume 2, pages 757–760, 1990.

[125] H. R. Berenji. Refinement of approximate reasoning-based controllers by reinforcement learning. In *Proceedings of the Eighth International Machine Learning Workshop*, pages 475–479, 1991.

[126] M. Berthold and K. Huber. *From Radial to Rectangular Basis Functions: A New Approach for Rule Learning from Large Datasets. Technical Report 15-95*, University of Karlsruhe, 1995.

[127] L. Bochereau and P. Bourgine. Extraction of semantic features and logical rules from a multilayer neural network. In *International Joint Conference on Neural Networks*, pages 579–582, 1990.

[128] A. E. Bryson and Y. C. Ho. *Applied Optimal Control*, Blaisdell, New York, 1969.

[129] W. Buntine and A. S. Weigend. Bayesian back-propagation. *Complex Systems*, 2:603–643, 1991.

[130] G. Carpenter and A. H. Tan. Rule extraction: from neural architecture to symbolic representation. *Connection Science*, 7(1):3–27, 1995.

[131] J. P. Cater. Successfully using peak learning rates of 10 (and greater) in back-propagation networks with the heuristic learning algorithm. In M. Caudill and C. Butler, editors, *IEEE First International Conference on Neural Networks (San Diego, 1987)*, volume 2, pages 645–651, IEEE, New York, 1987.

[132] A. Cauchy. Mèthode gènèral pour la resolution des systémes èquations simulationes, 1847.

[133] L. W. Chan. Efficacy of different learning algorithms of back-propagation networks. In *Proceedings IEEE TENCON-90*, 1990.

[134] L. W. Chan and F. Fallside. An adaptive training algorithm for back-propagation networks. In *Computer Speech and Language*, 2:205–218, 1987.

[135] F. L. Chung and T. Lee. A node pruning algorithm for back-propagation networks. *International Journal of Neural Systems*, 3(3):301–314, 1992.

[136] A. Cleeremans, D. Servan-Schreiber and J. L. McClelland. Finite state automata and simple recurrent networks. *Neural Computation*, 1(3):372–381, 1989.

[137] W. G. Cochran. *Sampling Techniques*. Wiley, New York, 1977.

[138] M. W. Craven and J. W. Shavlik. Using sampling and queries to extract rules from trained neural networks. In *Proceedings of the 11th International Conference on Machine Learning*, pages 37–45, 1994.

[139] C. Darken. Personal communication.

[140] J. Chang, C. Darken and J. Moody. Learning rate schedules for faster stochastic gradient search. In S. Y. Kung, F. Fallside, J. Sorensen and C. A. Kamm, editors, *Neural networks for Signal Processing 2, IEEE Workshop*, pages 3–13. IEEE, New York, 1992.

[141] J. Diederich. Explanation and artificial neural networks. *International Journal of Man–Machine Studies*, 37:335–357, 1992.

[142] L. Dillon, R. Hayward, J. Hogan and J. Diederich. Automated knowledge aquisition. *Technical Report*, 1996.

[143] S. E. Fahlman. Fast learning variations on backpropagation: An empirical study. In D. S. Touretzky, G. Hinton and T. Sejnowski, editors, *Proceedings of the 1988 Connectionist Models Summer School*, pages 38–51, Morgan Kauffmann, San Mateo, CA, 1989.

[144] B. Fälldin. SLAR, *Side Looking Airborne Radar—Signal Processing, Design and Evaluation*. Master's thesis, Chalmers Technical University, Gothenburg, 1993.

[145] V. V. Federov. *Theory of Optimal Experiments*. Academic, New York, 1972.

[146] R. Fletcher. *Practical Methods of Optimization*, volume 1, Wiley, New York, 1975.

[147] M. A. Franzini. Speech recognition with back-propagation. In D. S. Touretzky, G. Hinton and T. Sejnowski, editors, *Proceedings of the Ninth Annual Conference of the IEEE Engineering in Medicine and Biology Society*, pages 1702–1703, Boston, MA, 1987.

[148] P. Frasconi, G. Gori, M. Maggini and G. Soda. Unified integration of explicit knowledge and learning by example in recurrent networks. *IEEE Transactions on Knowledge and Data Engineering*, 7(2):340–346, 1995.

[149] L. M. Fu. Rule learning by searching on adapted nets. In *Proceedings of the Ninth National Conference on Artificial Intelligence*, pages 590–595, 1991.

[150] R. G. Gallager. *Information Theory and Reliable Communication.* Wiley, New York, 1968.

[151] S. Gallant. Connectionist expert systems. *Communications of the ACM*, 31(2):152–169, 1988.

[152] D. G. Garson. Interpreting neural-network connection weights. *AI Expert*, (April):47–51, 1991.

[153] S. Geva and J. Sitte J. Local response neural networks and fuzzy logic for control. In *Proceedings of the Second IEEE International Workshop on Emerging Technologies and Factory Automation*, pages 51–57, IEEE, New York, 1993.

[154] S. Geva and M. Orlowski. Simplifying the identification of decision rules with functional dependencies processing. *Technical Report QUT NRC* September 1996, Queensland Univerity, 1996.

[155] N. Gilbert. Explanation and dialogue. *The Knowledge Engineering Review*, 4(3):235–247, 1989.

[156] C. L. Giles and C. W. Omlin. Extraction, insertion and refinement of symbolic rules in dynamically driven recurrent networks. *Connection Science*, 5(3&4):307–328, 1993.

[157] C. L. Giles and C. W. Omlin. Rule refinement with recurrent neural networks. In *Proceedings of the IEEE International Conference on Neural Networks*, pages 801–806, IEEE, New York, 1993.

[158] L. Giles and C. Omlin. Rule revision with recurrent networks. *IEEE Transactions on Knowledge and Data Engineering*, 8(1):183, 1996.

[159] L. Goldstein. Mean square optimality in the continuous time Robbins Monro procedure. *Technical Report DRB-306*, Department of Mathematics, University of Southern California, 1987.

[160] M. Golea. On the complexity of rule extraction from neural networks and network querying. pages 51–59, 1996.

[161] W. Murray, P. E. Gill and M. H. Wright. *Practical Optimization.* Academic, London, 1981.

[162] H. Sawai P. Haffner, A. Waibel and K. Shikano. Fast back-propagation learning methods for neural networks in speech. *Technical Report*, ATR Interpreting Telephony Research Laboratories, 1988.

[163] S. K. Halgamuge and M. Glesner. Neural networks in designing fuzzy systems for real world applications. *Fuzzy Sets and Systems*, 65(1):1–12, 1994.

[164] S. J. Hanson and D. J. Burr. What connectionist models learn: Learning and representation in connectionist networks. *Behavioral and Brain Sciences*, (13):471–518, 1990.

[165] R. O. Harger. Object detection in clutter with learning maps. *SPIE Synthetic Aperture Radar*, 1630:176–186, 1992.

[166] M. R. Hestenes and S. Stiefel. *Methods of Conjugate Gradient for Solving Linear Systems. J. Res. National Bureau of Standards*, 49:409–436, 1952.

[167] R. C. Holte. Very simple classification rules perform well on most commonly used datasets. *Machine Learning*, 11:63–91, 1993.

[168] S. Horikawa, T. Furuhashi and Y. Uchikawa. On fuzzy modeling using fuzzy neural networks with the back-propagation algorithm. *IEEE Transactions on Neural Networks*, 3(5):801–806, 1992.

[169] R. H. Horn and C. A. Johnson. *Matrix Analysis.* Cambridge University Press, Cambridge, 1985.

[170] R. A. Jacobs. Increased Rates of Convergence Through Learning Rate Adaptation. *Neural Networks*, 1:295–307, 1988.

[171] J. A. Kinsella. Comparison and evaluation of variants of the conjugate gradient method for efficient learning in feed-forward neural networks with backward error propagation. *Network*, 3:27–35, 1992.

[172] H. F. Korth and A. Silberschatz. *Database Systems Concepts*. McGraw-Hill, New York, 1991.

[173] A. H. Kramer and A. Sangiovanni-Vicentelli. Efficient parallel learning algorithms for neural networks. In *Advances in Neural Information Processing Systems*, volume 1, pages 75–89, Morgan Kaufmann, San Mateo, CA, 1988.

[174] G. M. Kuhn and P. Herzberg. Some variations on training of recurrent networks. In *Proceedings of CAIP Neural Networks Workshop*, pages 15–17, Morgan Kaufmann, San Mateo, CA, 1990.

[175] K. J. Lang and M. Witbrock. Learning to tell two spirals apart. In G. Hinton, D. S. Touretzky and T. Sejnowski, editors, *Proceedings of the 1988 Connectionist Models Summer School*, pages 52–59, Morgan Kauffmann, San Mateo, CA, 1989.

[176] Y. Le Cun. Generalization and network design strategies. In F. Fogelmann Pfeifer, Z. Schleter and L. Steels, editors, *Connectionism in Perspective*, pages 52–59, Elsevier, Zurich, 1989.

[177] H. Leung and T. Lo. Chaotic radar signal processing over the sea. *IEEE Journal of Oceanic Engineering*, 18(3):287–295, July 1993.

[178] D. G. Luenberger. *Linear and Nonlinear Programming*. Addison-Wesley, Reading, MA, 1984.

[179] D. J. C. MacKay. Bayesian interpolation. *Neural Computation*, 4(3):415–447, 1991.

[180] D. J. C. MacKay. A practical Bayesian framework for back-prop networks. *Neural Computation*, 4(3):448–472, 1991.

[181] D. J. C. MacKay. Information-based objective functions for active data selection. *Neural Computation*, 4:590–604, 1992.

[182] M. Martinez, J. Juan, C. Casar, R. Jose and G. Miguel-Vela. A neural network approach to Doppler-based target classification. In *Proceedings of the IEEE International Radar Conference (RADAR 92)*, pages 450–453, IEEE, New York, 1992.

[183] R. Masuoka, N. Watanabe, A. Kawamura, Y. Owada and K. Asakawa. Neurofuzzy systems—fuzzy inference using a structured neural network. In *Proceedings of the International Conference on Fuzzy Logic and Neural Networks*, pages 173–177, 1990.

[184] S. Mitra. Fuzzy mlp based expert system for medical diagnosis. *Fuzzy Sets and Systems*, 65(2&3):285–296, 1994.

[185] M. F. Moller. *Cm Algoritmen*. Masters Thesis, *Technical Report, Daimi IR-95*, Computer Science Department, Aarhus University, 1990.

[186] M. F. Moller. Adaptive preconditioning of the Hessian matrix. In *Efficient Training of Feed-Forward Neural Networks*. PhD thesis, Computer Science Department, Aarhus University, 1993.

[187] M. F. Moller. Efficient training of feed-forward neural networks. *Technical Report Daimi PB-464*, Computer Science Department, Aarhus University, 1993.

[188] M. F. Moller. Exact calculation of the product of the Hessian matrix of feed-forward network error functions and a vector in $o(n)$ time. *Technical Report Daimi PB-432*, Computer Science Department, Aarhus University, 1993.

[189] M. F. Moller. A scaled conjugate gradient algorithm for fast supervised learning. *Neural Networks*, 6(4):525–533, 1993.

[190] M. F. Moller. Supervised learning on large redundant training sets. *International Journal of Neural Systems*, 4(1):15–25, 1993.

[191] J. Moody and C. J. Darken. Fast learning in networks of locally tuned processing units. *Neural Computation*, 1:281–294, 1989.

[192] J. D. Moore and W. R. Swartout. A reactive approach to explanation. In *Proceedings of the International Joint Conference on Artificial Intelligence*, pages 1504–1510, 1989.

[193] H. Okada, R. Masuoka and A. Kawamura. Knowledge based neural network—using fuzzy logic to initialise a multilayered neural network and interpret postlearning results. *Fujitsu Scientific and Technical Journal*, 29(3):217–226, 1993.

[194] J. R. Orlando, R. Mann and S. Haykin. Classification of sea-ice using a dual polarized radar. *IEEE Journal of Oceanic Engineering*, 15(3):228–237, July 1990.

[195] D. B. Parker. Learning logic. *Technical Report TR-47*, Center for Computational Research in Economics and Management Science, Massachusetts Institute of Technology, Cambridge, MA, 1985.

[196] Z. Pawlak. *Rough Sets—Theoretical Aspects of Reasoning about Data*. Kluwer, Deventer, 1991.

[197] B. A. Pearlmutter. Fast exact multiplication by the Hessian. *Neural Computation*, 7: 45–51, 1993.

[198] S. Nowlan D. Plaut and G. Hinton. Experiments on learning by back-propagation. *Technical Report CMU-CS-86-126*, Department of Computer Science, Carnegie Mellon University, Pittsburgh, PA, 1986.

[199] M. Plutowski, G. Cottrell and H. White. Learning mackey-glass from 25 examples, plus or minus 2. In Hanson, Giles and Cowan, editors, *Proceedings of Neural Information Processing Systems*, volume 4. Morgan Kauffman, San Mateo, CA, 1993.

[200] A. Ralston and P. Rabinowitz. *A First Course in Numerical Analysis*. McGraw-Hill, New York, 1978.

[201] H. Robbins and S. Munro. A stochastic approximation method. *Annals of Mathematics and Statistics*, 22:400–407, 1951.

[202] K. Saito and R. Nakano. Medical diagnostic expert system based on pdp model. In *Proceedings of IEEE International Conference on Neural Networks*, pages 255–262, 1988.

[203] D. Sanger. Contribution analysis: a technique for assigning responsibilities to hidden units in connectionist networks. *Connection Science*, 1:115–138, 1989.

[204] T. Sejnowski and C. Rosenberg. Parallel networks that learn to pronounce English text. *Complex Systems*, 1:145–168, 1987.

[205] S. Sestito and T. Dillon. *Automated Knowledge Acquisition*. Prentice-Hall, Englewood Cliffs, NJ, 1994.

[206] C. E. Shannon and W. Warren. *The Mathematical Theory of Communication*. University of Illinois Press, Urbana, IL, 1964.

[207] N. E. Sharkey and S. Jackson. An internal report for connectionists. In R. Sun and L. Bookman, editors, *Computational Architectures Integrating Neural and Symbolic Processes*, pages 223–244. Kluwer, Boston, MA, 1994.

[208] S. B. Thrun, J. Bala, E. Bloedorn, I. Bratko, B. Cestnik, J. Cheng, K. De-Jong, S. Dzeroski, S. E. Fahlman, D. Fisher, R. Hamann, K. Kaufman, S. Keller, I Kononenko, J. Kreuziger, R. S. Michalski, T. Mitchell, P. Pachowicz, Y. Reich, H. Vafaie, K. Van de Welde, W. Wenzel, J. Wnek and J. Zhang. The monk's problems: a performance comparison of different learning algorithms. *Technical Report CMU-CS-91-197*, Carnegie Mellon University, 1991.

[209] S. B. Thrun. Extracting provably correct rules from artificial neural networks. *Technical Report IAI-TR-93-5*, Institut fur Informatik III Universität Bonn, 1994.

[210] A. B. Tickle, M. Orlowski and J. Diederich. Dedec: decision detection by rule extraction from neural networks. *Technical Report QUT NRC* September 1994, Queensland University, 1994.

[211] T. Tollenaere. Supersab: Fast adaptive back-propagation with good scaling properties. *Neural Networks*, 3:561–573, 1990.

[212] B. A. Huberman, A. S. Weigend and D. E. Rumelhart. Predicting the future: A connectionist approach. *International Journal of Neural Systems*, 1:193–209, 1990.

[213] K. Plunkett, V. Marchman, and S. L. Knudsen. From rote learning to system building: acquiring verb morphology in children and connectionist nets. In D. S. Touretzky, J. L. Elman and G. E. Hinton, editors, *Connectionist Models: Proceedings of the 1990 Summer School*, pages 201–219. Morgan Kaufmann, San Mateo, CA, 1990.

[214] D. E. Rumelhart and D. Zipser. Feature discovery by competitive learning In D. E. Rumelhart, J. L. McClelland, and the PDP Research Group, editors, *Parallel Distributed Processing*, volume 1, pages 151–193, MIT Press, Cambridge, MA, 1986.

[215] J. Moody and C. Darken. Learning with localized receptive fields. In D. Touretzky, G. Hinton and T. Sejnowski, editors, *Proceedings of the 1988 Connectionist Models Summer School*, pages 133–143, Morgan Kaufman, San Mateo, CA, 1988.

[216] Y. Le Cun. Une Procedure d'Apprentissage pour Reseau a Seuil Assymetrique. In *Proceedings of Cognitiva*, 85:599–604, 1985.

Index

Printed and bound by CPI Group (UK) Ltd, Croydon, CR0 4YY

17/10/2024

01775685-0017